"This book shows the sizzling fut
enthusiasm to dispel the doom-and-gloom that paralyzes most
discussions of climate change."

Durwood Zaelke, President, Institute for Global Governance and Sustainable Development

"Today, solar is the fastest-growing source of renewable energy in America. That's not by fluke. Tam Hunt's new book offers important insights into solar energy's amazing success story—and why solar is so important to our nation's future."

Rhone Resch, President & CEO, Solar Energy Industries Association (the largest U.S. solar industry association)

"Of all renewable energy sources, solar is the most abundant and will play a major role in the conversion of the world's energy infrastructure to a clean and renewable one. Tam Hunt's book is easy to read and understand and is key for helping to educate the public and policymakers about solar's part in this energy revolution."

Prof. Mark Jacobson, Director of Stanford University's Atmosphere/Energy Program

"Tam's new book is a readable and engaging look at the inter-twined solar, electric vehicle and battery storage revolutions, from an insider with over a decade of experience in policy debates and battles in California, one of the centers for developments in these exciting new fields."

Terry Tamminen, Former Secretary, California Environmental Protection Agency, and current President of Seventh Generation Advisors.

Solar: Why Our Energy Future Is So Bright

Photography by Aramis Photography

ISBN: 978-1508786283

Alexander Jones
Christmas 2017
from Mom & Dad

I hope can play a role in our future sustainable economy!

"The future's so bright I gotta wear shades." Timbuk3.

Table of Contents

Foreword

The future of solar is amazingly bright. Over the past decade no energy industry has grown as fast in terms of the research base, the level of manufacturing, the number of new businesses, and the creativity of both end-users and new companies selling both on-grid and off-grid energy solar energy services. This dynamic and exciting theme is the central narrative of this high-energy overview of our energy system and where it's going.

Tam Hunt captures not only the incredible diversity of the solar energy field today, but also the degree to which solar energy can be the disruptive technology that opens doors for the energy system we want, not just the one we happened to inherit.

In this book Tam interweaves the story of the solar industry and its dramatic—game changing—evolution. He describes the ways that companies, normal homes, and even off-grid homesteads in Hawaii can all benefit, and have all shaped the industry. The interplay between market evolution in China, Germany, California, Italy, India, Kenya, and elsewhere, and manufacturing in an increasingly diverse set of countries, has changed the landscape of an entire industry.

Tam's story does not begin or end with the story of solar alone, and that is where solar moves from a good idea and a good reality, to a disruptive "black swan" that shows the power of new ideas. In addition to the dramatic expansion of solar energy—over 50% growth per year for over a decade—the solar revolution has enabled a new wave of thinking about energy more generally.

First, the old model of monolithic utilities has been called into question. Innovative new companies, and some innovative utilities are finding ways to take advantage of the two-way, real-time energy system that is now a possibility. Creative people and nimble companies are building on opportunities that solar energy, smart low-cost sensors, and the ideas that have inspired the "Internet of things" have all brought together. This new milieu is now coming together to permit the innovators to reinvent the largest industry on the planet: the energy industry.

The exciting thing about revolutions, of course, is the moment when all things seem possible, and creativity can take many directions. For Tam, the near future is one of escalating opportunities.

Clean energy is not about simply replacing dirty electrons with clean, green, electrons; it is about energy services that are simply better than what was there before. Instead of just greening the grid with solar, Tam's vision is of distributed solar energy as the tip of the iceberg, with electric vehicles and energy storage as the "apps" that go beyond green to redefine the global energy industry.

Instead of just cleaning the fuel in our cars (and improving urban air in particular, which brings benefits to people regardless of whether they own electric vehicles), Tam envisions more. In his vision of the future, clean, electric vehicles, become smart, driverless cars that free up commuters to work or relax during the commute.

We will see what additional innovations increasingly-cheap solar power brings us, but Tam's vision of the future is certainly as promising as any, and highlights beyond all else the ability of solar to enable a world of our choosing. *Solar: Why Our Energy Future Is So Bright* describes just that, and is a vision of the future you will greatly enjoy reading about.

Daniel M. Kammen,
Professor in the Energy and Resources Group and in the Goldman School of Public Policy;
Founding Director, Renewable and Appropriate Energy Laboratory
University of California, Berkeley

Preface

As I write these words in early 2015, oil prices have taken a tumble, falling more than half from their recent peak in mid-2014 at over $100 a barrel. Natural gas and coal prices have also tumbled in recent years. What, then, is the future of energy? Are the sky-high fossil fuel prices of not long ago a thing of the past? Are the rays of hope moving us away from dirty fossil fuels disappearing? Does our future hold a renewed and sustained abundance of cheap fossil fuels and a loss of confidence in renewable alternatives?

Don't count on it.

It is far more likely that our future will within two to three decades be powered mostly by renewable energy like solar, wind and biomass. Electric vehicles, many of them self-driving, will be on their way to ubiquity in the same timeframe and our roads and communities will be far cleaner, less noisy, and more enjoyable as a result.

The first half of this book focuses on the electric grid and how we produce electricity; the second half focuses mostly on oil and petroleum alternatives like electric vehicles. I also look at how you can best invest in the boom in renewables, considering the relevant risk factors, and employing an investor rather than a trader mentality.

Why am I so optimistic that our future will be primarily renewable? Haven't we seen this movie before, where an encouraging green shoot dies back, like in the late 1970s and early 1980s when solar power got its first big boost and then fell back? Well, the difference this time is that we have reached the point where growth in renewable energy is being driven primarily by economics alone and not by subsidies or mandates. We are not far from the point at which renewable energy explodes in what I call the "solar singularity."

The solar singularity occurs when solar becomes so cheap that it becomes the default new power source and more and more locations can affordably be covered in solar panels. Since the sun doesn't always shine, energy storage and other renewables like biomass and geothermal will help to make solar power reliable and we'll see a fully renewable energy grid in an increasing number of countries in the coming decades, joining the handful of countries that already have completely renewable grids.

Bloomberg New Energy Finance agreed with my optimism about renewables in a report from early 2015 with a title that says it all: "Seven reasons cheap oil can't stop renewables now: Oil is cheap. So is gas. Neither matters." "You couldn't kill solar now if you wanted to," said Jenny Chase, the lead solar analyst with Bloomberg New Energy Finance. Indeed. The solar singularity is nigh and I explain why in this book.

While solar continues its long boom, the U.S. economy and all global economies are becoming far more efficient. This is a natural progression we see in all countries as they develop: we learn how to do more with less energy because we learn how to use our energy more efficiently. We are also seeing most developed nations now declining in population, further reducing the energy required to run the modern world. 100 of the approximately 200 nations in the world are actually witnessing population decline. This is an ironic problem to deal with since the prevailing wisdom has been for some time that we are facing a population explosion. While most developing nations are still growing in population, these nations too will turn around before long and it's likely that the global population will be declining within a few decades. This decline will cause its own set of problems, in terms of economic growth and paying for health care, for example, but it will be a good thing in terms of our impact on the planet.

I discuss in these pages the big picture but also the "small picture" on the ground in terms of my personal experience with going solar at my little off-grid home in Hawaii and going all-electric with my first electric car purchase in 2014. A lot of my more detailed discussions focus on California because that is where I have done the most work as a renewable energy lawyer and policy expert. But California is often a harbinger of what is to come nationwide and even globally, due to our size and progressive attitude toward the environment, so these discussions are highly relevant to areas outside of California.

While this book is focused mostly on solar—and hence its title—it is really about four concurrent revolutions in the energy field: 1) the solar power revolution; 2) the energy storage revolution; 3) the electric vehicle revolution; 4) the driverless car revolution. These revolutions are of course intertwined and mutually dependent in various ways, but the end result of these revolutions as they transform societies each in their own way will be a far cleaner, cheaper and more convenient system for powering our homes and businesses as well as getting people and goods around.

I wrap some of my personal story into this book for two reasons: it will show the readers how I came to my conclusions, based on my background and expertise; and it will (hopefully) make for a more interesting read by injecting the human element into the narrative. There will be plenty of numbers and diagrams as we go (we're talking about energy technologies, after all) but I'll also share my own trials, tribulations and happy moments in the energy field over the last decade.

In short, this book will make you “energy literate” and capable of understanding the big trends in energy today, and also capable of capitalizing on these trends with smarter investments.

Tam Hunt, Santa Barbara, April 2015

Acknowledgments

This book is the end result of a dozen years of work in the renewable energy field, countless conversations, conferences, workshops, briefs, symposiums, articles, books, and private contemplation. As with anything in life it takes a wide network of contacts to learn anything worth learning and to bounce ideas around. I've had the benefit over the years of conversations, some brief, some long, with the following people who I owe some gratitude, in no particular order: Hank Hewitt, a long-time partner in crime as we have tried to show people why gravity is inevitable; Sahm White, for being diligent, brilliant and always patient; Craig Lewis, for being brilliant and cranky at the same time; Ted Ko, for being a true believer in renewable energy; Dan Kammen, for being a distant mentor in various ways; Dave Davis, for being a tough but fair boss back when I had a "real job"; Sigrid Wright, for helping me to write gooder; Michael Chiacos, for teaching me a thing or two about electric cars and solar; Greg Morris, for being a great client for many years and giving me generally free-rein to advocate for better state policies on electric vehicles on behalf of the Green Power Institute in Berkeley; to Eric Wesoff and Stephen Lacey for giving me a platform to spout my nonsense at GreenTechMedia.com, and the same to Jenn Runyon at RenewableEnergyWorld.com; and to my father, Christopher Hunt, for always being passionate about energy issues and buying me a copy of Vaclav Smil's *Energy at the Crossroads* way back when.

Energy terminology cheat sheet

Watt	Standard unit of electrical capacity (the ability to produce energy)
Kilowatt	1,000 watts. A home-sized solar system is usually 3-5 kilowatts. Similar to one horsepower, which is about 0.75 kilowatts
Kilowatt-hour	Standard unit of electrical energy. One hour's production of energy from a one-kilowatt capacity power source
Btu	British Thermal Unit, a common unit for measuring energy content of fossil fuels
Quad	One quadrillion btus, a common unit for comparing very large-scale energy use between countries or between economic sectors
Barrel of oil	53 gallons of oil, the common unit for oil production and shipments
EIA	The Energy Information Administration, the U.S. energy statistics agency
IEA	The International Energy Agency, the Western Nations' energy watchdog agency that works very closely with the EIA. I cite both of these agencies frequently on data and forecasts, not because they're necessarily right, but because they're mainstream.
AEO	The EIA's Annual Energy Outlook, a forecast of U.S. energy demand and production.
WEO	The IEA's World Energy Outlook. As the name suggests, a global energy forecast.

Chapter 1

Why is energy such a big deal?

Energy is a staple of the modern world, just behind air, food and water in terms of things we need to survive. Maybe most of us don't literally need energy other than food in order to survive, but to live in a modern economy we certainly do.

Where does our energy come from? We've been extracting energy from our environment for millennia, in the form of wood, peat or cow dung for our fires, in the form of water wheels and windmills, and in the last couple of hundred years in the form of fossil fuels. Coal was the first major fossil fuel to be utilized, starting in a serious way in the 17th Century in Europe, with England as the leading miner and consumer of coal. The Industrial Revolution, which began in England, was integrally related to the use of coal as a highly concentrated and readily available form of energy.

Oil was the next big fossil fuel. It was also discovered millennia ago and used for various non-energy purposes. But after its discovery in gusher form at a well at Titusville, Pennsylvania, in 1859, the Oil Age soon began. Oil became the transformational fuel of the 20th Century, allowing us to transport people and goods increasingly rapidly in cars, ships and airplanes. Oil was used for electricity production in the U.S. and many other countries for some time but today it is used almost exclusively for transportation fuel, plastics and chemical production.

Natural gas was the next big thing, with production of this lightweight fossil fuel ramping up in earnest in the 20th Century but really finding its stride in the last thirty years. Since 1980, global natural gas production has more than doubled. Natural gas is the Ron Burgundy of energy: kind of a big deal. It's become even a bigger deal with the advent of hydraulic fracturing (fracking) and horizontal drilling.

Natural gas extracted via fracking now constitutes about 40% of all natural gas produced in the U.S., up from almost nothing a decade ago. This is a remarkable increase, taking place mostly in the last five years, and made more remarkable when we consider that natural gas has now displaced King Coal as the largest source of fossil fuel primary energy production in the U.S.

According to the Energy Information Administration (EIA), natural gas accounted for about 31% of U.S. primary energy production in 2014 and is our largest single type of energy produced here in the U.S. Natural gas has surpassed coal production (about 26%), is almost twice the energy from oil production (about 17%), and is almost triple the energy we obtain from renewables (11%, including large hydro). The shale revolution has yet to reach countries outside of North America, however, with no other country producing more than 1% of its total gas from shale.

Where does renewable energy fit within this sea of fossil fuels? It is no secret that most renewable energy sources have been tiny blips for most of the industrial era. Industrialization and globalization have been driven by fossil fuels, and nuclear power to some degree in many countries. Hydropower has also been a very significant source of electricity since the mid-20th Century, with some dams reaching truly mega-scale. For example, the Grand Coulee Dam in Washington State, the biggest in the U.S., provides almost seven gigawatts of electricity at full power, enough for about seven million homes. This dam and a number of other large dams provide more than two-thirds of that state's power.

Four countries obtained all of their electricity from hydropower in 2008: Albania, Bhutan, Lesotho, and Paraguay. 15 other countries produce a majority of their electricity from hydropower, including, in no particular order: Venezuela, Iceland, Norway, and Sweden. China, however, is the new king of hydropower (and energy more generally), with a ton of enormous dams either already built or planned. Their biggest, the Three Gorges Dam, is more than three times the size of Grand Coulee Dam, at 22.5 gigawatts.

Hydropower has, however, fallen out of favor in many countries, at least in terms of new construction, because of the very large footprint and impact of these mega-projects. For example, in California, hydropower projects over 30 megawatts don't count as renewable energy for purposes of the state renewable energy mandate. Even though hydropower is technically a renewable source of energy, when I use the term renewable in this book I will be referring to all renewables other than large hydropower. I'll occasionally use the term "new renewables" also to refer to renewables other than large hydro, and specifically to refer to wind, solar and biomass, which are the real growth stories in the last couple of decades.

What is the status of non-hydro renewables in our emerging 21st century? Well, I'm glad you asked. The very short answer: they're still a very small portion of overall power consumption—solar, for example, is only about one percent of the global electricity mix in 2014, and less than one percent in the U.S. But at current rates of growth new renewables will be a very large portion of our energy mix in just a couple of decades. In fact, I argue in later chapters that one percent is actually *halfway* to market dominance. Keep reading for solutions to that head scratcher.

Some countries have grown the new renewable technologies of wind and solar remarkably rapidly. For example, Germany, examined in more detail in Chapter 5, obtains about one-third of its electricity mix from wind, solar and biomass, almost all of which has come online in the last fifteen years. For a country that is the fourth biggest economy in the world this is a remarkably rapid rate of change.

The International Energy Agency (IEA), the energy watchdog for the western nations, which I'll cite a lot in this book, surprised a lot of people with their announcement in 2014 that solar power could be the dominant global electricity source by 2050. This was a surprise because the IEA is known as a rather conservative and stodgy outfit, and has historically been far too pessimistic about the future growth of renewables. But they're clearly changing their tune as technology develops. IEA calculated that by 2050 we could see a 16% global share for solar photovoltaics ("PV panels," which are becoming ubiquitous on rooftops in many countries around the world) and an 11% share from concentrating solar power (CSP), which relies on various types of mirrors to concentrate sunlight onto a central receiver, for a total of 27% solar electricity by 2050. This wasn't a prediction by the IEA; rather, it's a scenario that IEA created to show what could come to pass under current trends for renewables.

Figure 1. *IEA modeling of global solar PV growth by 2050 (source: IEA Technology Roadmap: Solar Photovoltaic Energy, 2014 edition).*

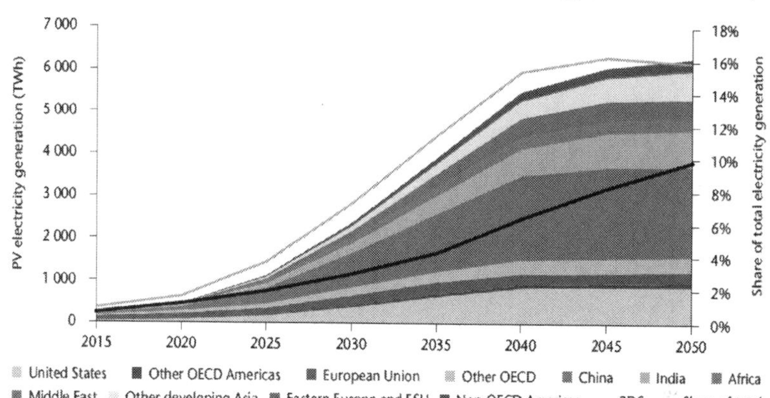

I flesh out in later chapters my reasoning for an even more aggressive growth projection for global solar than that provided by IEA. My arguments rest primarily on "learning curve" trends that shows a reliable relationship between increases in installed capacity and cost reductions. Increases in capacity lead to cost declines, which lead to increased capacity, which lead to..., in a virtuous circle that makes me feel warm and fuzzy about our future.

Even though new renewables are still a very small portion of the energy mix today, in the U.S. and around the world, it bears mentioning that most of our energy sources are various forms of solar power already. Fossil fuels are correctly viewed as a type of solar power because they are fossilized (hence the name) plant and animal matter that grew from the power of the sun. Wind power is also a type of solar because winds depend entirely on the sun's heat. Ditto with hydropower because the water cycle is driven entirely by the sun's heat. Geothermal and nuclear power are the only types of power we currently use that aren't forms of solar power. Geothermal is a type of nuclear power because it depends on heat generated from the radioactive processes in our planet's core. Geothermal is an excellent source of reliable renewable energy but its range is unfortunately quite limited based on current technologies. Efforts are underway on "enhanced geothermal" technologies that can access heat sources far deeper than is currently financially viable. When this happens we'll see geothermal take off too, but it's not at all clear when this will happen.

Energy is also a big deal because energy consumption is a strong predictor of wealth and per capita gross domestic product (GDP). Figure 2 shows energy consumption by country and a quick glance confirms that the richest nations generally are those with the highest per capita consumption.

Figure 2. *Energy consumption by country (source: International Energy Agency).*

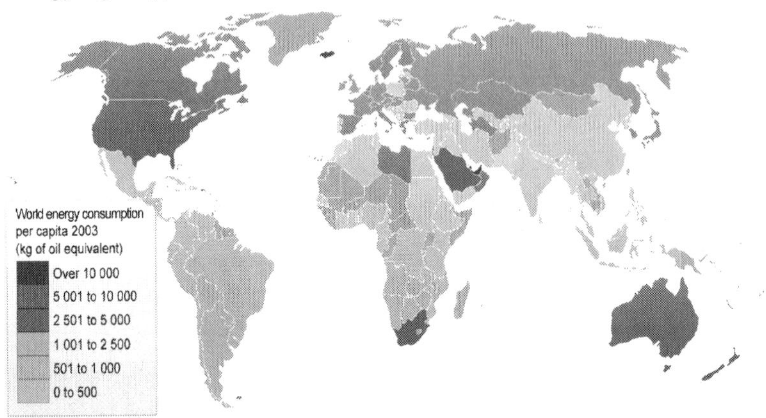

As nations grow, they use more energy, which helps them grow, in another (potentially) virtuous circle. Many nations, particularly in Europe, have been able to significantly reduce per capita energy use based on conservation and energy efficiency, while still enjoying strong growth and quality of life.

Energy use is, of course, also linked strongly to climate change. I'm not going to dwell much on climate change in this book because I don't need to. What? Well, my point is that what we're already seeing happen with renewables, electric cars, and improved energy efficiency is the right set of solutions when it comes to climate change anyway. I stopped debating climate science years ago and I'm better off for it. As we'll see in later chapters, there is a very compelling case for the renewable energy transformation even if we ignore the perils of climate change. The IEA—again, a historically stodgy outfit—turned even more heads with another 2014 report that projected a $71 trillion (with a "t") net savings through renewable energy development and other climate change mitigation measures around the world by 2050. Why such large savings? The lion's share comes from fuel cost savings from increased efficiency and fuel shifting to electricity away from petroleum.

IEA has created a nifty website for data visualizations here (www.iea.org/etp/explore), which allows you to play around with the various projections and see some cool graphics.

The good news on climate change, then, which isn't talked about too much yet, is that the current trends in renewable energy and electrification of transportation will yield global solutions to climate change in the coming decades. And because the renewable energy transformation will yield very large net financial savings, along with job creation and many other benefits, it's clear that we should do this even if (gasp) current climate change science is totally wrong. Climate change is not, then, a "dispositive" factor in our argument for renewables, to use a useful legal term (I've got to show somehow that I was paying attention in law school).

Whether the coming renewable energy transformation will do enough to reduce atmospheric greenhouse gas concentrations in the time-scale required to avoid major climate problems is an open question. I am not, then, arguing that we can forget about climate change because of today's renewable energy growth rates. To the contrary, what I am saying is that the renewable energy transformation headed our way, and in particular the "solar singularity" just a few years away, will likely be just the ticket required for very serious climate change mitigation, even if we can't avoid all the consequences of the greenhouse gas emissions we've already dumped into the atmosphere over the last couple of centuries. As with all things, time will tell if I'm right.

In short, energy is a big deal because we wouldn't have the modern world without it. And if we keep on going with a fossil fuel-based energy system we probably won't have a modern world either. So this book is about the good news on energy and how we're rapidly leaving fossil fuels behind.

Chapter 2

Where we're going and how to get there

I like data. There, I said it. Actually, I *love* data, so this chapter will contain quite a few numbers and charts looking at how we can make the transition to a predominantly renewable energy economy here in the U.S. There are many reasons to make this transition but I'll take as a given that we should make this transition and focus on how we can best do this.

For those who are allergic to numbers put this book down now. All right, you kept reading. Now let me sum up the conclusions here in the beginning of this chapter. If we are to be successful in mitigating climate change and achieving a sustainable and independent energy system, we need to ride the waves already coming our way and do our best to start new waves where we have the power to do so. As already mentioned, riding these waves will also lead to substantial cost savings for countries around the world.

The biggest wave, by far, which is already underneath us and swelling, is solar power. We need to ride this wave as far as it will go—and it will go far. The cost of solar power has plummeted in the last few years by over 50% and we are seeing solar power costs at or below the cost of utility power in an increasing number of jurisdictions already; this is generally known as "grid parity." A 2014 report found that Germany, Italy and Spain are now at grid parity for solar PV and many other countries are close. Deutsche Bank went even further in a report in early 2015 and predicted that 80% of the world would enjoy grid parity by 2017. A big part of why solar is growing so fast is the dramatic cost reductions we've seen:

over 80% in the last five years or so. First Solar, one of the biggest solar panel manufacturers and solar farm developers, predicted in early 2015 that the all-in cost for now solar projects would be as little as $1 per watt by 2017. When we compare this to the comparable cost for utility-scale solar projects from just a few years ago—well over $5—we can see why solar is becoming such a compelling proposition.

I call the next big step for solar after grid parity, the point at which solar power becomes the default new power source in a majority of jurisdictions around the world, the "solar singularity." When this moment is reached, solar power will take off and become the dominant power source relatively quickly. Chapter 3 focuses more on the solar singularity. My feeling is that we'll see solar reach half or more of our power supply in the U.S. sometime in the late 2030s or early 2040s. I explain this projection later in the book. That's still a ways off but it's pretty soon in terms of the time normally required for major energy transitions.

It seems a little crazy at first blush to say this, but we are already effectively at the halfway point to solar ubiquity because we reached 1% of new power plant installations from solar in 2013 (and about 1% of total installed capacity in 2014). This is strange but *mathematically* true because 1% is halfway in terms of the doublings required to get from nothing to 1% and from 1% to 100%. So in terms of the time required we may indeed be halfway to solar dominance. This is an example of Kurzweil's Law of Accelerating Returns. (Chapter 7 examines this argument in more detail).

The next big wave is energy storage. It's nowhere near as certain as the solar wave because it's a newer technology and its swell is only being dimly felt for now. But with the right policy support, and an army of smart entrepreneurs, this energy storage wave will be just as rideable as the solar wave. Germany is leading the way (again) on installations, with about 13,000 residential PV+battery systems expected to be installed in 2015, already over 1,000 a month. One large Germany battery manufacturer expects this figure to reach 10,000 a month in the next five years as costs come down by about 5% per year.

California is arguably leading the way on the utility-scale side (see Chapter 15 for more on California's energy storage efforts). Energy storage will be key for integrating variable renewables like solar and wind into our grids as penetration reaches high levels (we won't need it for a number of years but by planning now for when we do need it we will make the transition that much more smooth).

Energy storage is more useful for the grid than natural gas backup power because energy storage devices can go both ways: they can absorb *and* dispatch power to the grid whereas natural gas plants can only dispatch. So it's two for the price of one when it comes to energy storage. The price, however, is the catch for now: even though good research suggests that energy storage may already be cost-effective when we properly account for the benefits to the grid, most observers would agree that there's still a lot of room for energy storage costs to come down. My feeling, admittedly tinged with some hope, is that we'll see the same trend in the energy storage business that we've seen in solar in the last five years, with strong demand prompting a huge ramp in production and thus big drops in price, which is what we see in tons of different fields in what is known as the "learning curve" phenomenon.

The last wave I'll mention here is the energy efficiency wave. We use energy very wastefully because, frankly, energy is still really cheap—even here in California where I live most of the time. We in the U.S. waste well over half of all the energy that is actually available in our system (Figure 1). And when we consider the potential for conservation—behavior change—to reduce energy use even further, we could, it seems, be just as productive as we are today on an energy budget half or more of what we currently use.

For the western half of the U.S., a grid consisting of large amounts of solar and wind, hydro and biomass, and backed up with energy storage and flexible natural gas plants, could readily provide all or a large part of the power we need to maintain and grow a modern economy. Other parts of the U.S., particularly the south, don't have quite the renewable energy endowment that the western U.S. has, or even the Northeast. However, high voltage DC power lines are an option for areas without an abundance of renewable energy. While distributed energy and localized grids are to be preferred, I'd rather see renewables supply the South from power lines from Texas and the Midwest, or from large offshore wind farms in the Atlantic, than see the South continue with its coal-dominated power mix in perpetuity.

Now let's talk turkey

Ok, let's talk numbers. I'm going to describe a *plausible* pathway for the U.S. to become a predominantly renewable energy economy by 2035 to 2040. I stress "plausible" because of course this kind of forecasting is generally an exercise in futility due to so many unpredictable variables. But that shouldn't stop us from trying. I'll describe what the U.S. currently uses for electricity, transportation and other types of energy today and then project forward to 2040. Then I'm going to show in broad terms how we could reasonably make the transition to a predominantly renewable energy system by 2035-2040.

I'm going to use "quads" as my common unit, which is short for quadrillion btus (British thermal units; for some reason the acronym is not capitalized). Quads are commonly used when talking about large-scale energy use, such as in the context of entire nations' energy budget. For example, the U.S. used about 97 quads of energy in 2013 (Figure 1).

Figure 1. *U.S. energy flow chart 2013 (source: Lawrence Livermore National Laboratory).*

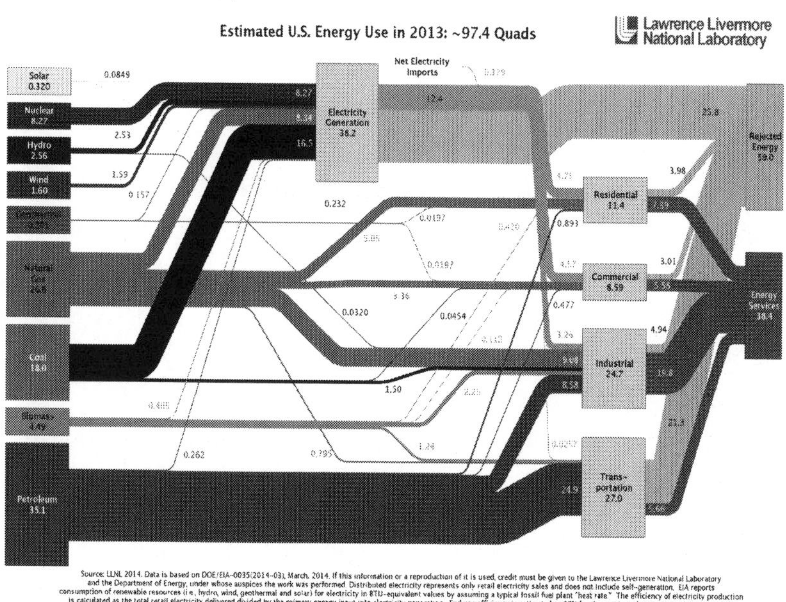

We can follow Figure 1 and group the various end-uses of energy into four sectors, which we'll tackle below: 1) transportation; 2) industrial; 3) commercial; 4) residential.

We see from this same chart that petroleum is the single largest energy source in the U.S., followed by natural gas and then coal. Renewable energy is still a relatively tiny share of the whole, even if we include large hydro, which isn't even considered renewable in some jurisdictions, like California, due to its environmentally-damaging impact.

Another major take-home from this chart—often called a "spaghetti chart" or, officially, a Sankey diagram—is the pronounced inefficiency in our energy use. 59 quads are "rejected," that is, wasted. Only 38.4 quads are used productively. This poses a big opportunity to use energy more efficiently.

Looking ahead, the Energy Information Administration projects a thirty percent increase in vehicle miles traveled by 2040 (EIA Annual Energy Outlook 2014). However, projected increases in fuel economy more than offset this increase in driving and the net result in EIA's forecast is actually a decline in transportation fuel demand by 2040, from 26.7 quads in 2012 to 25.5 in 2040.

EIA projects that electricity consumption will grow 0.9% per year through 2040 in its Reference Case, and natural gas consumption grows by about 0.8% per year through 2040, with industrial use of natural gas in chemical production showing the strongest growth.

In sum, EIA expects U.S. energy consumption to increase to about 106 quads by 2040, from 97 quads in 2013, an increase of 9%.

I'm going to define a "predominantly renewable energy economy" as 80% or more renewable energy, which includes renewable electricity, biogas and biofuels. No, it doesn't include nuclear. This makes the magic number to reach by 2040 or sooner about 85 quads (80% of 106 quads).

So how do we get to 85 quads from renewables?

Price-induced conservation and energy efficiency

First, we disagree with EIA about the likely trajectory of U.S. petroleum demand. Based in part on the arguments I set forth Chapter 16, I believe there is a good case for lower U.S. oil production than EIA projects, in the longer term, consequently higher prices, and a significantly lower energy content of produced barrels because unconventional oils contain less energy per barrel than conventional oil. The main points I make in that chapter concern: 1) much higher than normal decline rates for unconventional oil wells than for conventional oil wells and 2) a failure to consider the global impact of declining net oil exports as major oil exporters use ever larger portions of oil that they produce while their oil production declines over time.

EIA already includes in its forecasts an increasingly efficient economy because this has been the long-term trend. EIA projects a 2% decline in energy intensity per year through 2040. This means that our economy will produce the same goods and services each year with 2% less energy.

Figure 2. *EIA's energy intensity projections (source: EIA AEO 2014).*

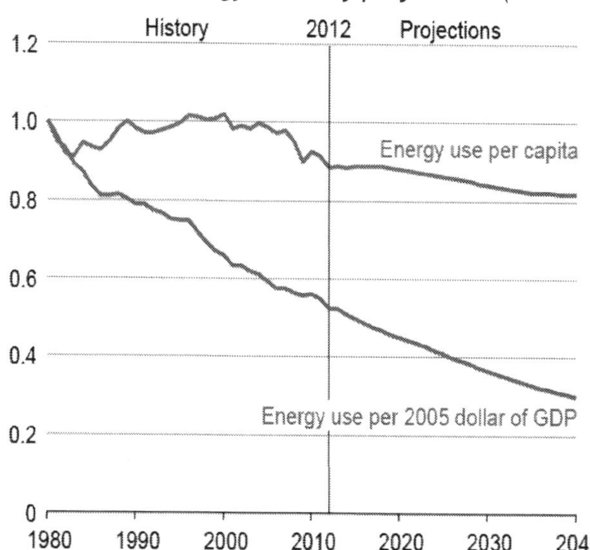

However, due to the factors mentioned above we may see substantially more price-induced energy conservation than EIA projects by 2040. Moreover, as we shift to electricity as a major transportation fuel a large increase in efficiency is achieved because electric vehicles are about three times more efficient than conventional cars in converting energy into motion. I'm going to, accordingly, assume that 25% of the 85 quads will be met by additional conservation and efficiency that is not included in EIA's current forecast. This brings us to 64 quads as the magic number to achieve by 2040.

Transportation energy

Let's start with the hard part first. Shifting to a predominantly renewable electricity system seems almost inevitable at this point. The transportation sector is a tougher nut to crack because we're still so dependent on petroleum. Electrification is the key, however, to weaning us from petroleum as well as coal and natural gas. There are many other ways to reduce petroleum demand—hybrid cars, smaller and more efficient cars, biking, walking more, carpooling, increasing busing and train routes, smarter urban planning, etc.—but to actually get us off petroleum we should look primarily to electrification of vehicles. I won't re-hash the arguments here, but I cover the major debate points over electric vehicles vs. fuel cell vehicles in Chapter 12. In sum, I don't see fuel cell vehicles as a significant player in our future.

EIA, as mentioned, projects 25.5 quads for transportation energy by 2040, but this includes only a small amount of electrification. Based on the logic described above for increased price-induced conservation and improved efficiency, I reduce this figure by 25% to 19.1 quads. This means that higher petroleum prices will induce a stronger shift away from traditional vehicles, and away from driving more generally, than EIA currently projects.

While biofuels like ethanol and biodiesel are far from perfect solutions, we can't ignore that they have in fact grown rapidly in recent years and are probably here to stay. Ethanol now provides about one million barrels of fuel per day in the U.S., which after adjusting for energy content, is equivalent to about 700,000 barrels per day of oil, or about 3.7% of U.S. consumption. Assuming only that biofuels production stays constant, subtracting this amount from 19.1 gives 18.4 quads for transportation energy needs by 2040.

Since I've defined a renewable energy economy as one that gets 80% or more of its power from renewables, we can reduce this 18.4 quads to 14.7 quads, which assumes that 20% of transportation will still come from fossil fuels by 2040. Since this chapter is just an outline I'm going to forecast at this point that this 14.7 quads will come entirely from electricity due to relatively rapid electrification of transportation through various types of electric vehicles. This is, of course, highly debatable and uncertain, but, again, I think it's plausible given the trends we're seeing today. We have 26 years to get there.

The astounding thing about electric vehicles is that they use energy about three times more efficiently than internal combustion engine vehicles. So switching our fleet to EVs entirely by 2040 would allow us to meet all 80% of our transportation needs with only about 5 quads of electricity. If you don't believe me double-check my math.

Electrifying our transportation sector also allows us to focus on how we produce electricity in our country as *the key task* for transitioning off fossil fuels economy-wide, which we discuss next.

Our renewable energy future

We currently produce a plurality of our electricity from coal, but an increasing share comes from natural gas and renewables. Nuclear power's share is significant but slowly shrinking. In 2013, the U.S. obtained 41% electricity from coal (down from over 50% just a few years ago), 26% from natural gas, 19% from nuclear, 13% from renewables (which includes large hydro), and 1% from petroleum.

Figure 3. *Sources of U.S. electricity production, trillion kilowatt-hours (source: EIA AEO 2014).*

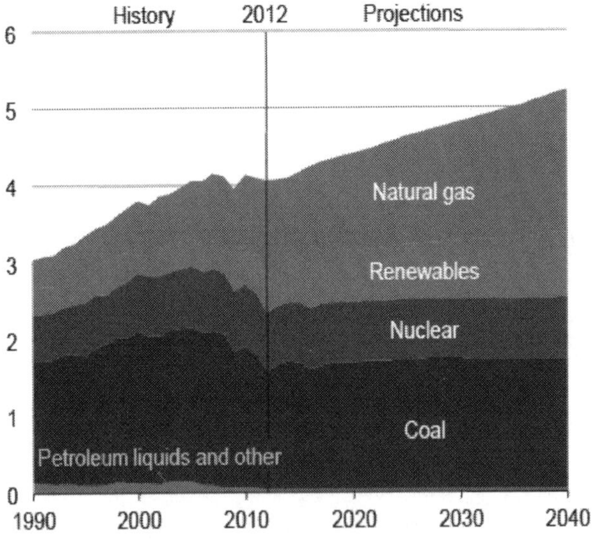

By 2040, EIA projects that the mix will change to 32% coal, 35% natural gas, and 16% each for nuclear and renewables. I think these projections are way off, due largely to the within-reach "solar singularity" mentioned above.

EIA has often been wrong when it comes to projecting renewable energy growth and also wrong when it comes to projecting fossil fuel production. In the case of renewables, they've almost always been too pessimistic and in the case of fossil fuels generally too optimistic. (The last few years have, however, shifted EIA's record on fossil fuels because the shale gas and oil revolutions have surprised almost everyone).

Anyway, using quads, our common unit, we convert EIA's numbers and get a revised forecast of 12.6 quads of electricity consumption in the U.S. by 2040, which includes my standard 25% reduction due to additional price-induced conservation and efficiency. Further reducing by 20%, allowing for our 80% definition of a "predominantly renewable energy economy" brings that down to 10 quads. But we need to add in our shifted transportation electricity demand (5 quads) and this brings our total renewable electricity goal to 15 quads.

I'm going to be bold and project, based on today's growth rates of solar and the almost-here grid parity for solar around the country, that we'll see solar grow to 50% of *all* electricity supply by 2035-2040, including our shifted transportation energy demand. This isn't just a wild guess. It's based on a projected growth rate of 30% per year from 2013 onward.

30% is actually a significantly lower growth rate than we've seen in the last five years (average 54% growth rate in solar electricity produced) but we should expect the rate to slow down over time since this is an almost universal pattern for the diffusion of new technology.

The 30% growth figure is admittedly a guess, but it's an educated guess based on the many positive trends in the solar industry and the fact that solar is so scalable and modular. Some areas of

Australia have reached 25% or higher residential penetration rates for solar in just a few years—the highest in the world—and there's no reason to think that we can't in the U.S. achieve similarly fast penetration once the tipping point has been reached. The U.S. only has about 0.3% solar penetration in 2014, but with steady growth rates this small base grows rapidly, spurred by low cost and massive scalability.

Wind power growth is less certain, for a variety of reasons, including less widespread wind resources than for solar, and additional permitting hurdles for wind turbines when compared to flat solar panels. Also, the annual growth rate of wind in recent years has been highly variable. Average growth for wind power in the U.S. over the last five years has been 25%. If we project only half that annual growth from 2013 onward (12.5%), we get to about 10 quads of wind power by the late 2030s. This leaves wiggle room, as with solar, for lower growth rates to still meet the projections for 80% renewables. In a surprising shift upwards from previous scenarios the Department of Energy released a new study in 2015 showing a "viable scenario" (short of making this a forecast) that wind could achieve 20% of U.S. electricity by 2030 and 35% by 2050.

Under these projected growth rates, wind and solar could bring us to about 80% of all electricity consumed by 2035 to 2040, leaving 20% to be provided by other renewables and/or natural gas and storage, and possibly some residual nuclear plants. The 15 quads of total electricity demand by 2040, including the shifted transportation energy, could, then, come from wind and solar backed up with battery storage, baseload renewables and residual conventional power sources.

Biomass and geothermal are important renewable energy technologies, particularly because they're generally baseload sources, i.e., they can produce power when needed. Biomass, geothermal and small hydro could probably provide the 20% remaining electricity needs by 2040, which would allow elimination of all fossil fuels in the electricity sector. Additional incentives will likely be needed for these technologies because they're currently not growing very fast, for a variety of reasons. Given the technical potential, however, for these resources, and given the many examples around the world of how smart incentives can bring new technologies to scale, there is a good argument for additional incentives for these baseload renewables.

Will we really eliminate fossil fuels in electricity generation by 2040? Almost certainly not. But that's not my point here. My point is to show that we *could* if we decide we want to, based on plausible growth rates for renewables in the coming years.

Maintaining grid reliability

Grid reliability is a major issue when it comes to high penetration of renewables. As discussed briefly in my introduction, we will need large amounts of energy storage to balance a predominantly renewable energy grid. For present purposes, I'm simply going to assume that the nascent energy storage wave I discussed above swells fast enough to allow integration of renewables at the levels I project in this article. This is a very big assumption and time will tell if I'm way off. Keep in mind, however, that I have allowed 20% of electricity to come from non-renewable sources in my definition of a renewable energy economy and this allows some padding to help balance a high renewables grid, along with lots of energy storage and interconnected grids for further balancing.

Substituting for natural gas

So far we've covered petroleum and electricity. This covers the lion's share of energy use in the four energy consumption sectors described previously: industrial, commercial, residential and transportation energy. It leaves out, however, heating, cooling and industrial use of natural gas. EIA calculates about 20.8 quads of natural gas use for heating, cooling and industrial processes by 2040. Reducing by 20% due to our definition of "predominantly renewable" we get 16.6 quads that we need to source from additional renewables to get to our goal.

Solar PV's poor cousin is solar water heating technology. In fact, solar water heating may be more prevalent in the world today than solar PV; it's just not as sexy. China solar water heating rivals the rest of the world's installed PV capacity, with about 118 gigawatts equivalent of solar thermal installed in China by 2010 and significant growth since then.

Solar water heating is growing in the U.S., but not as fast as solar PV. California has had a rebate program for solar water heating for a number of years, but it's still a relatively small program. A 2007 NREL study found only 0.5 quads of technical potential for SWH in the U.S. This leaves 16.1 quads to make up still. This is a tough sector to source predominantly from renewables, but for present purposes I'm going to project that a mix of SWH, biomass and renewable electricity can meet these 16.1 quads. This will require a higher growth rate for renewable electricity sources than I have projected above, so it may be the case that natural gas for industrial processes and heating and cooling will take a bit longer to source predominantly from renewables than transportation and electricity.

Cross-checking with others

Mark Jacobson and his team at Stanford have completed a huge amount of work in this area. His team published a draft 2014 paper looking at achieving a 100% renewable energy economy by 2050 and this paper provides good support for my projections here, though they don't see things changing quite as quickly as I do.

Jacobson's work has been incorporated into a very useful website with nice infographics showing how each state can achieve the transition. Here's the graphic for California and all the other 49 states (http://thesolutionsproject.org/infographic/), showing about 50% solar and about 35% wind by 2050, and a 44% improvement in energy efficiency.

The National Renewable Energy Laboratory also completed a major study in 2012 looking at what it would take to get to 80% renewable electricity by 2050 in the U.S. They project that wind will be the largest single source of electricity and they also show a higher percentage of biomass than I've calculated here. I think NREL is probably wrong about wind being the largest source of electricity by 2050, due to the growth rates of solar that we've seen in recent years, but I'll personally be entirely happy if wind does in fact stay the larger power source. Wind and solar are both great sustainable and cost-effective power sources.

I haven't discussed costs at all yet. However, it is clear already that transitioning to a fully renewable economy will *save* tons of money on a net basis. This is counter-intuitive to most. Aren't renewables more expensive? Well, historically they often have been, but that's changed a lot and the costs of renewables continue to fall while the cost of fossil fuels generally continues to climb. This means that after the costs of installation and maintenance are accounted for, the savings from zero fuel costs (for most renewables) and a far more efficient economy more than outweigh the costs. Jacobson has crunched the numbers and he and his team project a net savings of $4,500 per year per person in the U.S. When you factor in health benefits from far less pollution the savings almost double.

Another study from the International Energy Agency projected a net savings of $71 trillion (yes, with a "t") by 2050 resulting from the investments in new energy technologies required to keep the globe's temperature from rising more than two degrees Celsius. Again, these cost savings result primarily from fuel cost reductions.

We have, then, very good arguments that we should shift to a renewable economy even absent any climate or energy independence benefits—we could and should do it entirely as an economic boost. (*See* Chapter 20 for more on the debate about the costs of the transition to a renewable energy future; I duke it out amicably with Prof. David Victor on these issues in that chapter).

Final thoughts

It is worth reminding readers that the scenario I've sketched should be considered a high petroleum price scenario, based on the factors I've outlined above that relate to global oil production, net exports and net energy content. As such, if for whatever reason oil prices remain at current levels or lower it's very likely that my projections will be off. As with all things, I could be entirely wrong. We must keep in mind, however, the oft-quoted truth: the Stone Age didn't end because we ran out of stones. And the Oil Age is not going to end because we run out of oil.

Chapter 3

The Solar Singularity is nigh

"It's tough to make predictions, especially about the future," quipped Yogi Berra. I keep his wise admonition in mind as I make predictions about our energy future, but we have many reasons for optimism when it comes to the future growth of solar.

Here's the summary: solar is taking over. We can now see many years into the future when it comes to energy. And that future is primarily solar-powered. Why my optimism? Well, let me explain.

The "solar singularity" will, as mentioned in the last chapter, occur when solar prices become so cheap that solar becomes the default power source based on cost alone. We aren't there yet but we're probably just a few years away from that point, particularly since we're seeing energy storage costs declining strongly already, and we'll need a lot of storage to help firm up solar power as solar reaches higher penetration levels. (I'm not going to address storage in this chapter further but, of course, a grid can't run on variable solar power alone so we'll need storage and other backup technologies to ensure reliable grids as solar power penetration grows).

Why the "solar singularity"? This is a play on the "singularity" concept made popular by Ray Kurzweil and others. Kurzweil's 2005 book, *The Singularity is Near*, made a big impression on me when I read it in 2007. The book is all about the ability of exponential growth to lead to radical change far more quickly than we expect because we are inclined to think in linear terms rather than exponential terms.

A great example of this kind of thinking is the ancient Indian story of a king challenged to a chess game by a traveling sage. Before the game was finished the sage asked for a reward if he won the game. Confident in his ability to win, the king agreed to whatever the sage requested. The sage requested a modest prize: the king was to place a single grain of rice on the first square of the chessboard and double it with each successive square. The king didn't bat an eye before agreeing to this surprisingly modest prize. Inevitably the king lost and as his courtiers counted out the number of rice grains required to pay the same they quickly realized that it would take far more rice than existed not only in his kingdom, but in all of India and in all the world. The total is actually two to the 63rd power (the first square has only one grain of rice), which equals over 18 sextillion (18 followed by 18 zeros).

The point is that exponential growth gets very big very quickly. And this is why Kurzweil used the term singularity for this kind of growth.

Swanson's Law, named after the founder of SunPower, is a particular type of "experience curve" or "learning curve" that is an example of technology improvement in circumstances involving exponential growth over a sustained period. Swanson's Law states that the price of solar panels generally drops by 20 percent with every doubling of shipped panels. This has been the general trend since solar became a viable technology—hence its designation as a "law" even though there are times when some deviations from the trend take place. For example, from the mid-1990s until 2008 solar costs declined by relatively little, primarily due to stubbornly high silicon prices in a backdrop of increasing commodity prices across many markets, until the crash of 2008. Since 2008, however, panel cost declines have accelerated and the general trend is now back and then some.

When we compare recent cost declines for solar to other energy prices we get a pretty picture indeed and this is why solar is now getting very serious attention by investors and pundits alike.

Swanson provided some interesting additional insights into the nature of his eponymous law in a 2014 email in response to an article of mine:

> ["Swanson's Law"] is not really a "law" but merely expresses the fact that manufactured goods tend to follow a straight line on a log-log plot of price versus cumulative unit volume. To the best of my knowledge no one knows why this is so. It is curious that it usually continues to be so, even when the costs become largely commodity items like glass and steel. This has led some in the past to assume that the cost reductions would flatten out when commodity material costs dominate. In fact this has not been observed, in part because people find ways to use less materials, and because materials themselves follow a learning curve.

IEA releases an annual report on the global status of solar energy. Their 2014 report showed a phenomenal 39 percent growth in solar power, with 39 gigawatts added in 2013. The 2015 report showed about the same level of added capacity in 2014, bringing the total by the end of 2014 to about 180 gigawatts. It must be satisfying for Swanson to see his predictions come true in spades. When he wrote his 2006 paper, global solar installations were only about 5 gigawatts. We are now, in early 2015, at about 200 gigawatts, about forty times the installations in 2006, with prices declining much as he predicted.

Figure 1. *Global solar power growth through 2014 (source: IEA 2015 PVS report).*

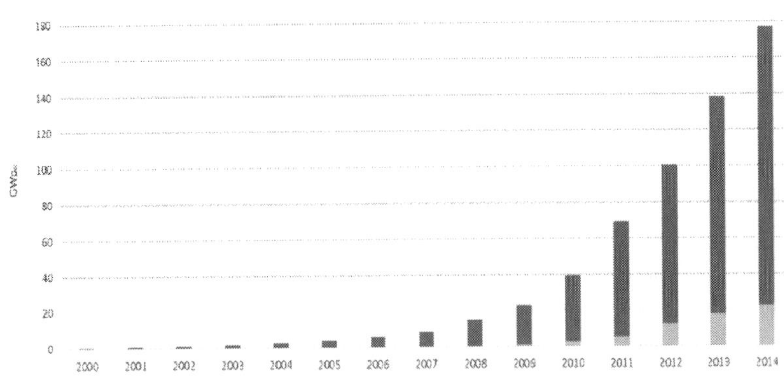

Total U.S. installations now stand at about 20 gigawatts, or 10 percent of the total and enough power for about four million U.S. homes.

What about subsidies?

Subsidies have been a big part of getting solar to where it is today but subsidies are becoming increasingly unnecessary as solar prices plummet. This is additional good news. California's residential and commercial solar rebate program (the California Solar Initiative or CSI) is all but gone as the rebates have been used up, and yet California's retail solar market is still growing strongly.

On the wholesale side, the federal 30% investment tax credit (ITC) is set to decline to 10% at the beginning of 2017. The conventional wisdom is that we'll see a big drop in installations when this happens. However, a silver lining to the Republican-controlled Congress and their antipathy to green power is that there is little hope at this point that the 30% ITC will be extended.

This means, contrary to the similar discussion with respect to wind power's tax credits over the last decade (they've expired a number of times, leading to a slowdown in installations for a year and then a rebound when the credit is renewed), there won't be a slowdown in anticipation of an eventual renewal of the tax credit. We should see solar companies simply adjust to the lower tax benefit and keep on trucking.

SunPower, a major player in today's markets, is already predicting little impact from the reduced ITC, based on the ability to develop profitable projects even with the reduced ITC of 10%.

James Smith, an investment analyst at Catapult Research, recently issued a very bullish report on solar, providing some good corroboration of my predictions here. He stated in his report: "I'm saying that if the cost of solar drops 20% in price every time the installed base doubles, it is only a matter of time before solar takes over from fossil fuels. My best guess is that it starts to really happen from 2017 onwards."

Is the past a reliable guide to the future?

Making predictions (especially about the future) is difficult because there is no guarantee, of course, that the past is a reliable guide to the future. However, when it comes to solar power we see the very clear trend of price reductions continuing for some time because there are no inherent limits to further reductions. Jeremy Rifkin has made the case that solar panels will become practically free with zero marginal cost for production, in his book *The Zero Marginal Cost Society*. As we'll see below, this is a reasonable prediction.

Solar panels are not the only cost component for solar systems and they are increasingly becoming a minor cost because of ongoing panel cost reductions. The main components of overall costs are now soft costs like labor and the "balance of system" costs for equipment like inverters, racks and wiring. However, these other costs are also declining substantially and groups like GTM Research predict further major cost reductions.

The basis for my predictions is, however, quite simple: we have reached the point where low costs are driving installations higher, which in turn drives costs lower, which in turn drives installations higher.... The virtuous circle seems to be locked in and based on history we can expect further 20% cost reductions with each doubling of capacity, with no inherent limit to cost reductions over time.

Under this trend, we can expect by 2020, under a 30% global rate of growth, to see total solar costs for utility-scale systems at around $0.84/watt, based on GTM Research's projected $1.10/watt for 2017. By 2025, the cost drops to about $0.54/watt and by 2030 it will be a practically free cost of $0.34/watt. By 2040, we can expect under these trends to see costs at about 14 cents per watt. A five kilowatt home-size system costs at this price only $700.

That counts as free in my book because that system will provide power for about 25 years at almost no cost above the initial installation cost. 25 years of production for $700 equates to about 2.8 cents per kilowatt-hour. For comparison, the average retail cost of power in California today is about 15 cents per kilowatt hour, so this future cost of solar power will be less than 1/5th the cost of today's power. And this analysis leaves out inflation. If we include inflation the comparison is far more favorable.

What could derail the solar singularity?

While I'm fairly confident in the coming solar singularity I'd be foolish not to recognize some inherent uncertainties about making such predictions. I'll discuss a couple of the biggest uncertainties here.

The biggest source of uncertainty is the rate of growth in installations. In my calculations above I assumed a 30 percent rate of growth (the same as that I used in the last chapter), which is reasonable given the far higher rates of growth we've seen in recent years (this results in approximately a 2.3-year doubling time). However, it is likely that we'll see growth rates decline for a variety of reasons. If installations increase at only 20% per year we see about $0.54/watt by 2030 and 28 c/watt by 2040. At only 10% growth we see about $0.75/watt by 2030 and $0.56/watt by 2040. At these price and installation levels the singularity still arrives but it's delayed.

Figure 4. *Projected cost declines based on Swanson's Law, ignoring inflation (source: Tam Hunt).*

	30% R.O.G.		20% ROG		15% ROG		10% ROG	
	Cost/kW		Cost/kW		Cost/kW		Cost/kW	
2017	$	1.10	$	1.10	$	1.10	$	1.10
2018	$	1.01	$	1.04	$	1.05	$	1.07
2019	$	0.92	$	0.98	$	1.01	$	1.04
2020	$	0.84	$	0.92	$	0.96	$	1.01
2021	$	0.77	$	0.87	$	0.92	$	0.98
2022	$	0.70	$	0.82	$	0.88	$	0.95
2023	$	0.64	$	0.77	$	0.85	$	0.92
2024	$	0.59	$	0.73	$	0.81	$	0.90
2025	$	0.54	$	0.69	$	0.77	$	0.87
2026	$	0.49	$	0.65	$	0.74	$	0.85
2027	$	0.45	$	0.61	$	0.71	$	0.82
2028	$	0.41	$	0.58	$	0.68	$	0.80
2029	$	0.38	$	0.54	$	0.65	$	0.78
2030	$	0.34	$	0.51	$	0.62	$	0.75
2031	$	0.31	$	0.48	$	0.60	$	0.73
2032	$	0.29	$	0.46	$	0.57	$	0.71
2033	$	0.26	$	0.43	$	0.55	$	0.69
2034	$	0.24	$	0.40	$	0.52	$	0.67
2035	$	0.22	$	0.38	$	0.50	$	0.65
2036	$	0.20	$	0.36	$	0.48	$	0.63
2037	$	0.18	$	0.34	$	0.46	$	0.62
2038	$	0.17	$	0.32	$	0.44	$	0.60
2039	$	0.15	$	0.30	$	0.42	$	0.58
2040	$	0.14	$	0.28	$	0.40	$	0.56

The second biggest source of uncertainty is the degree to which there are fundamental limitations in how fast power generation fleets can turn over. Most power generation assets are financed (amortized) over the course of many years and these investments often require long power sales contracts to justify such investments. This means that a lot of the fleet is locked in contractually at any given time. If a ton of solar is installed in any particular grid system the threat of "stranded costs"—costs that are at risk of not being recovered due to under-utilization or an early shut down—becomes high.

We'll see how the stranded cost issue shakes out in each country but there is good reason to believe that even if some grids see a slow-down in solar installations because of concerns about stranded costs, or other problems, that other countries will take up the slack and the general global trend of ever-increasing solar will continue apace.

One issue that I don't think will be a real problem in the next couple of decades is lack of space for new solar. For all practical purposes, the space for installing solar around the world is infinite. We'll run out of power demand long before we'll run out of space for solar. As costs plummet for solar, more and more countries will see solar become economically viable and more and more locations, such as roadways, areas over metro rail lines, etc., will be covered in solar.

In sum, we have some very good reasons to believe that the solar singularity is indeed nigh.

Chapter 4

Why electricity will save us

Ben Franklin famously stated about his scientific work on electricity: “If there is no other Use discover'd of Electricity, this, however, is something considerable, that it may help to make a vain Man humble.” Electricity is a relatively recent type of energy for us humans. While we have known about lightning, electric eels and other forms of electricity for millennia, electricity wasn’t firmly part of modern science until the formulation by Faraday and Maxwell in the Nineteenth Century of what are now known as Maxwell’s equations of electromagnetism.

With the advent of the light bulb in the mid-18th Century, the use of electricity quickly spread in the industrialized world. The first building to be lit entirely by electric light bulbs was the Savoy Theatre in London in 1881. The electric telegraph had arrived a few decades earlier but it wasn’t a significant source of electric consumption. Electric motors and radio became the next significant uses after the light bulb. Beginning in the last years of the Nineteenth Century electricity generation started to become a big industry. It’s an interesting historical fact that the first cars were originally electric but soon lost out to gasoline cars in the first two decades of the 20th Century due primarily to the longer range made possible by the advent of petroleum as a widespread fuel.

The history of the personal automobile is a good case study of why electricity is such a good energy carrier. The rise of the internal combustion engine, and petroleum as the fuel source, took place largely because petroleum became so plentiful in the U.S. in the early decades of the 20th Century. As we’ll see in later chapters, the first U.S. commercial production of oil took place in Titusville, Pennsylvania, in 1859. The U.S. became the largest producer of oil in the 1920s, surpassing Russia. From then until 1970, the U.S. was the largest oil producer in the world—a fifty-year reign that cemented the U.S. role as most powerful superpower for some time after our indigenous energy production peaked. U.S. oil production peaked in 1970 and declined to almost half its peak by 2008, at which point it started growing again with the advent of unconventional oil production. At this time of this writing it seems that we may see oil production return to its 1970 peak or even surpass this level. This is a big surprise because the 40-year trend was almost inexorably downward. But the fracking and unconventional oil revolutions have allowed what will very likely be a temporary major spike in oil production that will probably run out of steam in a few years.

When oil gushed from the ground and the U.S. was awash in the stuff in the early part of the 20th Century, it didn't make sense to worry about energy efficiency. Nor was anyone worried about climate change or energy independence when the U.S. produced so much oil and before we realized that human activities can impact climate. Nor did we worry at that time about the air pollution caused by burning oil or other fossil fuels. It took the Cuyahoga River in Ohio to catch on fire in 1969 (and a few times before then too) to spark the environmental movement. By 1973, the major federal environmental laws were in place, courtesy of the Republican president Richard Nixon—demonstrating that environmental issues can and should be a bipartisan concern.

We are now, in the first years of the 21st Century, increasingly concerned about energy efficiency, air pollution, energy independence and climate change. All of these factors weigh heavily in favor of electrification of our energy system and moving away from petroleum and other fossil fuels. Yes, oil and other fossil fuel prices are relatively low here in 2015, but it is all but certain that they will return to previous highs in the coming years. Oil was above $100 a barrel for most of the period from 2006 to 2014 and it's likely that we'll be back in that range before too long. Ironically, however, we may see "peak demand" before "peak oil," as discussed in Chapter 16. Nevertheless, it is likely that we'll see prices go far higher again rather than a "new normal" of stable and low prices. The one certainty in energy markets seems to be volatility.

Back to electricity and here's the key point of this chapter: Electricity will save us because it's a very efficient energy carrier and can eventually be sourced entirely from renewable and non-polluting sources like solar, wind, geothermal, biomass and small hydro. The best example for the innate energy efficiency of electricity is comparing the electric vehicle to the internal combustion vehicle (ICE, today's typical car). The electric vehicle uses energy about three times more efficiently than the ICE (the federal website www.fueleconomy.gov states that EVs convert 59-62% of the energy from the grid into to power at the wheels versus 17-21% for internal combustion vehicle energy conversion). As we'll see in Chapter 9 EVs also use electricity about 2.5 times more efficiently than hydrogen fuel cell vehicles, due to the conversion losses in creating hydrogen with electricity and then converting hydrogen back into electricity in the fuel cell.

This difference in energy efficiency means that the same amount of fossil fuel energy would take an EV 300 miles for every 100 miles the ICE would go. This very substantial difference is primarily due to heat losses that are inherent to any combustion process. This is why car engines get hot: a lot of energy is spread out into the environment rather than being used to move the car and its passengers. EVs use electric motors, which rely on electricity to turn a magnetic coil, which turns the drive shaft. So there are very few moving parts in EVs when compared to ICEs, which have hundreds or thousands of moving parts. The "moving parts" in EVs, other than the motor and the drive shaft and wheels are the tiny electrons moving from the battery to the motor. This relative lack of moving parts translates into far lower maintenance requirements for EVs. For example, my Fiat 500e, an all-electric car that I first leased in 2014, will not require maintenance for the first 20,000 miles of my ownership. A new ICE car requires maintenance every 5,000 miles.

Electricity may correctly be described as a carrier of energy rather than as a primary source of energy because all electricity that we use in our devices has to be produced from some primary source of energy such as solar, wind, biomass, hydro, coal, natural gas or nuclear power, to name a few. But with the solar singularity arriving before too long, and with the current global electric grid doing pretty well at providing electricity where it is needed (except, of course, in the many developing countries that are still energy-impoverished), we can rest assured that electricity is going to be with us for some time regardless of the primary form of energy behind it.

Because electricity is such a useful form/carrier of energy I will generally use kWh (kilowatt-hours) or other increments of watt-hours (megawatt-hours, gigawatt-hours, etc.) as a common unit of energy from here on in this book. I used the quad unit in Chapter 2 when looking at the U.S. as a whole because this is the common convention and is very convenient when looking at very large energy uses. From now on, I'll use varieties of watt-hour units.

Chapter 5

Germany's energy transition: a great example for the world to follow

Germany, the world's fourth largest economy, has for over a decade now led the world in terms of its commitment to wean itself from fossil fuels and reduce its greenhouse gas emissions. This program is known as the *Energiewende*, or "energy transition," a program made into law in 2011. The key laws behind the Energiewende were put in place a decade earlier.

This commitment to renewables has not been merely rhetorical. Germany has installed more solar power than any other country, despite having a rather modest solar resource, and was until fairly recently also the world leader in installed wind (China and the U.S. have since surpassed Germany on solar). Figure 1 shows Germany's installed renewable electricity through 2013.

Figure 1. *Germany's installed renewables.* (Source)

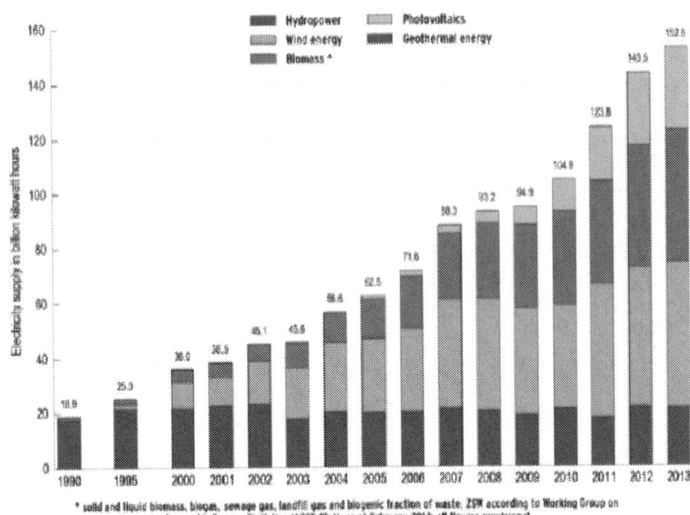

Renewable energy is now producing about 30% of all of Germany's electricity needs. On some days, solar power provides half of the total grid demand, which is particularly remarkable considering that Germany only started installing solar a bit more than a decade ago. Total installed solar capacity now exceeds 38 GW. Installed wind power is over 35 GW and biomass power still produces almost twice as much power than solar in Germany even though we hear far less about this third major renewable, which is also baseload (reliable) power.

Germany installed a remarkable 22 GW of solar from 2010-2012 (about 7 GW each year), before slowing to about half the annual rate in 2013 and slowing further in 2014 as previous reforms went into place. Becoming a victim of its own success, in 2014 Germany's legislature, the Reichstag, with the support of Prime Minister Angela Merkel, enacted additional reforms of the longstanding *Energiewende*, designed to both slow down the torrid rate of growth and to change the pricing system for new contracts.

Numerous reports in U.S. and other English-language media have announced the death of Germany's commitment to renewable energy and greenhouse gas emissions reductions. Fox News commentators proclaim this shift with a glint in their eye, happy that Germany's radical green policies have been shown to be wrong and Germany is now being forced to backtrack.

But is any of this true? It is true that reforms were enacted in 2014, but they're evolutionary rather than revolutionary, and it is not at all true that Germany is backtracking on its long-term commitments. Let's examine the evidence. First, a little history.

The origins of the Energiewende

The primary law behind the transition is known as the EEG (an acronym for a long German title), which was first put in place in 2000. The EEG is a feed-in tariff, modeled on a similar U.S. law passed in 1978 (PURPA, discussed further in Chapter 8). A feed-in tariff allows producers to sell power to the utility under a long-term contract at a set price. It has been key to Germany's success because it provided the necessary certainty for major investments to be made.

The EEG has been amended many times, but most substantially in 2012 and 2014. The Heinrich Böll Foundation Energiewende website includes a brief summary of its history:

> The German Energiewende did not just come about in 2011. It is rooted in the anti-nuclear movement of the 70s and brings together both conservatives and conservationists – from environmentalists to the church. The shock of the oil crisis and the meltdown in Chernobyl lead to the search for alternatives – and the invention of feed-in tariffs.

The EEG put in place a system of feed-in tariffs for various technologies, with prices set administratively at a level deemed necessary to spur development.

The prices have been adjusted downward (never up) numerous times since the beginning, a mechanism designed to spur early development and rapidly declining costs. Figure 2 shows this history for the solar feed-in tariff for systems 10 kW and below (rates are lower for larger systems and much lower for other technologies like wind and biomass). There was a 74% feed-in tariff price decline from 2004 to 2013, which, as the chart shows, correlates very well with the total system price (the green squares).

Figure 2. *Feed-in tariff price reductions for solar 10 kW and below (source: Lawrence Berkeley National Lab)*

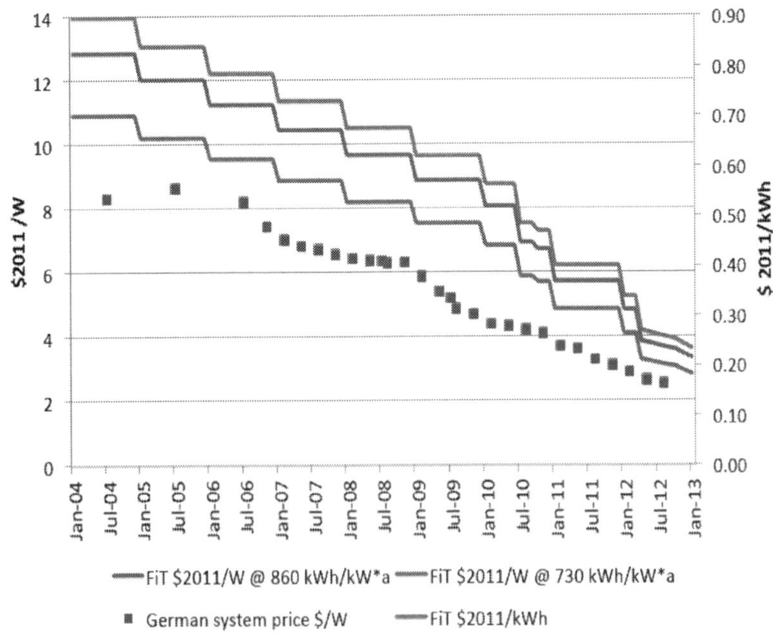

"EEG 2.0"

The EEG was substantially changed in July 2014 with strong support for the changes in the German Parliament. The new changes took effect August 1.

The key changes were as follows:

- New solar contracts will generally be limited to 2.5 gigawatts (GW) per year, down from as much as 7 GW each year in 2011 and 2012 (more contracts can be signed but incentives drop as this target is exceeded)
- New onshore wind contracts generally limited to 2.5 GW per year
- New biomass contracts limited to 100 MW per year
- A pilot program will begin in 2017 to test a tendering/auction system as an alternative to the feed-in tariff, but the pilot will only apply to ground-mount solar projects at first
- The goal of 80% renewable electricity by 2050 was maintained, as well as intermediate goals of 40-45% by 2025 and 55-60% by 2035

The tentative steps toward an auction system and away from feed-in tariffs, for larger systems (over 500 kW), is perhaps the biggest change, but as Paulos points out it's actually not a big change from the system in place today, under which wind power projects already have to sell their power on the open market and then receive a "market bonus" to ensure that the market rate they receive is equivalent to the posted FIT price.

The difference with the planned 2017 pilot for large ground-mounted solar is that it will use the auction/tendering process to determine the appropriate FIT equivalent price, rather than setting the price administratively. Accordingly, the big change here seems to be a shift to a market-based price-setting process instead of an administratively-set price.

Before then, new solar systems will receive the FIT price but under a different mechanism than before, also relying on the market premium concept. So, in effect, the FIT is still in place, but in concept it's been revised in anticipation of probably bigger changes down the road in 2016, when "EEG 3.0" is planned.

A key feature of FITs, the constant price during the 20-year contract, is maintained under the new system. So the market certainty that this constant price provides will still be operative.

Restrictions on the size of the market went into place previously for 2014, with the current annual "corridor" for solar planned at 2.5-3.5 GW. Again, this slowdown in annual installations was a product of the amazing success of the FIT, not a sign of failure. Germany's farmers and companies have simply been adding new renewables, particularly solar, faster than ever expected and fast enough that Germany's grid operators and others have convinced the policymakers to slow down the transition a bit. This is the result both of an overly rapid growth in renewables and fears that the transition may be costing too much.

Perhaps the most telling fact with respect to the costs and the politics of the Energiewende is that almost everyone in Germany still supports the Energiewende. A poll in February 2014 found that fully 89% of Germans supported the Energiewende. This is a strong rebuttal to some rather breathless and sensationalized reporting in the U.S. about the alleged failures of the Energiewende. For example, this New York Times article talked about a new "energy poverty" taking hold in Germany due to the Energiewende. The facts don't support this article's slant.

What has been the effect of Germany's Energiewende around the world?

Correlation is not causation but very strong correlations are often good indications of causation. We can look at the coincident timing of Germany's FIT, plummeting prices of solar that track the declining FIT price, and the rise of, in particular, China's solar production capacity, and draw a pretty clear picture of why solar prices came down so far so fast. In one sentence: Germany's robust and long-term FIT policies created huge demand for solar and prompted countries around the world, but particularly China, to ramp up production to meet demand. The end result was almost an 80% or more module price decline in just a few years, in line with Swanson's Law, as we'd expect because Germany was generally the world's biggest market from 2005 through 2013.

A final word is warranted on the geopolitics and energy politics prompted by Russia's new assertiveness in the region. Specifically, Russia's interactions with Ukraine and Crimea and its demonstrated willingness to cut the natural gas spigot to major customers have rightly worried Germany. If Germany is to transition to renewables and also enhance its energy independence it may need to eventually wean itself not only from coal and nuclear but also from Russian natural gas. About 1/3 of Germany's natural gas currently comes from Russia but Germany imports fully 90% of its total natural gas needs (including for electricity and for heat, the lion's share of natural gas consumption). These figures highlight the difficulty that Germany will have in weaning itself from fossil fuels by 2050. My feeling is, however, that the new issues with Russia will, if anything, accelerate Germany's push on the Energiewende. There is already some evidence for this and I expect to see more on this in the coming years.

Summing up, EEG 2.0 is a substantial reform of the EEG law but it does not diminish Germany's long-standing commitment to the energy transition. There is a clear shift in the 2014 reforms away from the FIT model for all renewables and toward a more market-based model for larger renewables. This is, in my view, regrettable because Germany's policies to date have succeeded in bringing new renewables below the cost of grid power, and in many cases below the cost of wholesale power, so new FIT contracts can increasingly save ratepayers money.

The certainty of the FIT process, both in terms of a long-term fixed price contract, and knowing the price ahead of time, have been major factors in spurring Germany's remarkable transformation. Germany's far-sighted policies have been successful in bringing costs down dramatically, to the point where ratepayers can actually save money on new contracts, so why pull back on this policy now that is has achieved this major milestone? While the new reforms will water down the major features of the former FIT policy somewhat, it appears that Germany is still well on its way to achieving its energy transition goals, and probably long before the current deadlines.

Chapter 6

How solar can help the developing world leapfrog the dirty energy phase

Some commentators have criticized the focus on renewable energy and other "green" technologies in part because they fear that an undue focus on low emissions and non-fossil fuel sources of energy will doom many developing nations to energy poverty for a long time to come.

The idea is that these countries would be better off economically from using fossil fuels like coal and natural gas to build their own modern grid systems with large central-station power plants, which has been the default model for many decades. Even though these plants may exact some environmental costs, it is thought, a modern central-station grid will allow these countries to develop more quickly than would otherwise be possible, and they can eventually enter the developed-nation club due to the benefits that a modern grid can provide.

I appreciate the concerns behind these arguments but I think they're wrong in many, if not most, situations. As with all of my analysis, even though I'm making conclusions that I will defend I always come at this kind of issue objectively and I won't ignore facts that don't support my conclusions. In other words, I'm always going to do my best to be fair with the facts.

Jim Rogers was CEO of Duke Energy, based in North Carolina and the largest electric holding company in North America, for a number of years. He retired in 2014 to focus on scholarship and philanthropy. One of his initiatives is an effort to bring renewable energy to the developing world. He states eloquently in a 2015 article in EnergyBiz magazine:

> My interest in the lack of access to electricity in many countries of the world began with a chance meeting with a young man in a Kenyan village. He was holding a cellphone in the middle of nowhere, with not a power line in sight.
>
> "How do you charge that thing?" I asked.
>
> "I walk three hours to the charging station," he said.
>
> Wow, I thought. He walks three hours to the charging station - six hours in one day - to charge his cell phone. I can barely stand it when I check into a hotel and find there's not an outlet conveniently placed next to my bed.
>
> I've spent most of my career providing electricity to millions of people, and I'm stunned by the global statistics: [more than] one in six people worldwide lack access to electrical power. That means [1.5] billion people have no Internet, no water pumps, no bright lights to study by. Around another billion and a half people or so have limited access. There's no question that electricity is the foundation for economic development, education, women's rights, health and efficient farming. Let's give these people the chance to get ahead and take better care of their families. It is a human imperative. I believe that together we can make access to clean and sustainable electricity a basic human right.

Rogers is putting his money and time where his mouth is. His Global BrightLight Foundation has distributed free of charge, or sometimes in return for community service, over 70,000 combination solar LED lanterns and cell-phone chargers to families in Rwanda, Uganda, Zambia, Nepal, Peru, Bolivia, Haiti and Guatemala. After their initial efforts focused on the free distribution model they began charging for this product and are now turning profits from these sales back into increased sales around the world. The lantern that Global BrightLight distributes is available on Amazon.com for $20 for the mid-range model and $45 for the top model; both have great reviews.

A smaller and waterproof solar LED lighting solution is available from Luci for $15. I've used this model myself when camping and it provides great light over many hours on a single change. It's not that good for lighting up anything bigger than a tent or small room, but it can provide ambient light for a lot of purposes. It's also compressible, very lightweight, and entirely water and windproof.

These prices are about the same as a cell phone and service plan for many in developing countries and various options for payments, including microcredit programs, have proven quite successful in many circumstances. Many of those benefiting from these solar LED and cellphone-charging devices would otherwise spend the same or more on candles or kerosene lanterns and fuel, or have to walk hours to charge their cellphones. There is, thus, a major net benefit from going solar in these situations.

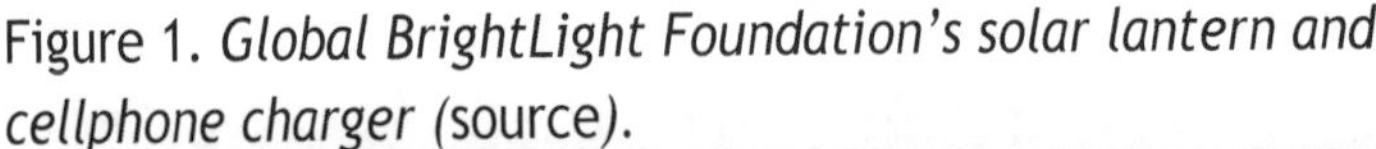

Figure 1. *Global BrightLight Foundation's solar lantern and cellphone charger* (source).

The debate over central-station generation (from fossil fuels) and a distributed renewable energy system is generally, in the case of the poorest countries, a non-debate: the central-station model clearly hasn't worked, more than 100 years after this model was created in Europe and the U.S., for the 1.5 billion people still without access to basic electricity services (according to the International Energy Agency's 2013 Pico PV report). And we are now seeing a number of companies and non-governmental organizations (NGOs) like Rogers' step into the gap left by the modern central-station model. Another 1.7 billion people don't have access to clean or reliable electricity because the available grid is so bad. This means that over 45% of the world's seven billion population could currently benefit greatly from a distributed renewable energy model for at least basic electricity needs like lighting, phone charging and small electronics use. To those without access to these basic features of a modern society such benefits can be literally life-changing—as Rogers'

simple example of a man walking six hours a day just to charge his cell phone makes clear.

"Pico PV" refers to very small-scale solar systems like those used in the U.S. to power an automatic gate or a traffic light, but increasingly used in the developing world as the primary source of power. The IEA's 2013 Pico PV report adds:

> Solar pico PV systems have experienced significant development in the last few years, combining the use of very efficient lights (mostly LEDs) with sophisticated charge controllers and efficient batteries. With a small PV panel of only a few watts essential services can be provided, such as lighting, phone charging and powering a radio. Expandable solar pico systems have entered the market. Households can start by buying a small kit, later adding an extra kit, allowing extra lights and services to be connected and even a small TV to be considered.

This warms my heart to know that people who crave access to the benefits of modern society can have many of these benefits while also avoiding some of the biggest downsides of that same society: the pollution, dependency, and landscape-scarring effects of the developed world's fossil fuel addiction.

IEA provides a strong warning, however: pico PV is not a full solution and it may have some downsides in terms of the "solar trap" of believing a community's energy needs are fully met with pico PV when this is, instead, only a partial move toward electrification and all the benefits that can bring:

> The majority of the 1.5 billion people mentioned above will have no grid connection for years to come, perhaps never, and for them solar pico PV systems can help in providing a few

> essential energy services. But it should also be realized that rural inhabitants usually prefer a grid connection, in order to watch colour TV and to iron their clothes when they want. Despite the provision of this initial level of service with pico solar PV systems, they should still be considered non-electrified and the so-called "solar trap" should be avoided.

Pico PV is, then, one step on the way toward electrification. This doesn't mean that pico PV is a step toward central-station generation. Rather, it should mean that the next step is building out more robust community-scale and home-scale PV systems, with good batteries, that can provide more robust power to these communities when needed. This larger-scale, but still distributed, solar and storage grid, is still far cheaper and faster to build than the traditional central-station fossil fuel grid that is considered the default option for modern development.

An example of this next step in solar development can be found in Tanzania in south-central Africa. OffGrid Electric, with major investments from the large U.S. solar company SolarCity, is investing millions of U.S. dollars to solarize that country. The products aren't that different than the solar lanterns offered by Global BrightLight but the objective is to install panels and lights in homes as a bit more permanent solution than the ultra-portable Sun King lights.

Solar Sister is a similar effort in Tanzania, but led by local partners, with U.S. Agency for International Development backing, as part of the $9 billion Power Africa initiative. Solar Sister has since 2009 brought solar power to almost 200,000 people in Tanzania, providing a major economic and health boost to this country of 50 million people.

What about intermediate-level economies?

The debate between central-station grids and distributed renewable energy grids is more significant when we look up the economic ladder of nations a little. Coal and natural gas costs can be very low at times, but these prices are always volatile. Large power grids can work quite well and in those countries, such as India, Brazil, Indonesia, Thailand, etc., where there are semi-functioning grids it is not as clear that going solar and distributed is a cheaper option at this time.

However, when we factor in the health costs of fossil fuel power, the lack of reliability of such grids, and the always-declining cost curves for solar power, it is very clear that even if fossil fuels can sometimes be cheaper than the solar alternatives today it is only a matter of a few years until solar will become the default choice in these countries too. This is the primary message of the "solar singularity": on the current solar development trajectory around the world we're going to see solar prices continue to plummet in the coming years.

In sum, solar power and a distributed grid offer substantial benefits to most developing countries today and these benefits will become more pronounced over time as the solar singularity works its magic.

Chapter 7

Why 1% is halfway to complete energy transformation

Solar power has changed in just a decade from an interesting pipe dream for greenie types to a mainstream source of power, due to its increasing cost-effectiveness in many countries around the world. This is a remarkable evolution and demonstrates well why CitiGroup recently stated that the “age of renewables” is upon us.

We’ve already seen how solar power is on a tear in previous chapters. Cumulative solar photovoltaic electricity production is now at one percent of total global electricity production, at about 200 gigawatts installed capacity.

Only 1%, you say? That’s tiny. Right. But that 1% is actually *halfway* to the goal of market saturation. Yes, 1% is halfway toward market dominance and I’ll explain why below.

Figure 1. *Global solar PV capacity, annual and cumulative (source: IEA).*

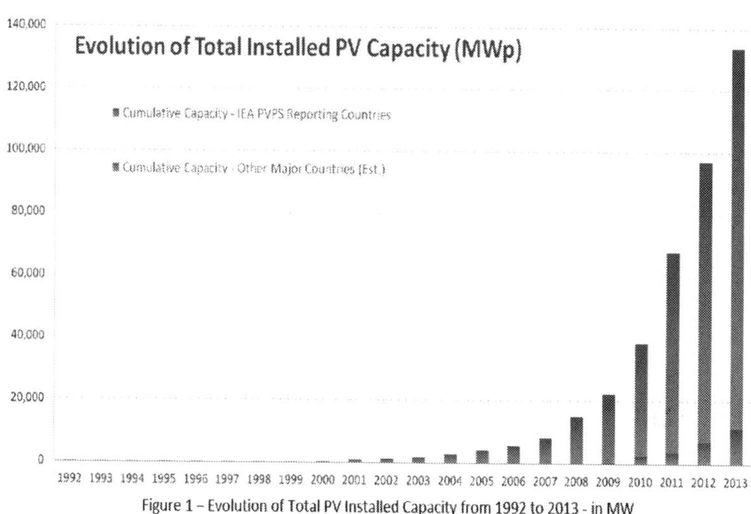

Figure 1 – Evolution of Total PV Installed Capacity from 1992 to 2013 - in MW

Ray Kurzweil, an American inventor and entrepreneur, discovered the Law of Accelerating Returns by studying numerous technology learning and adoption curves. The classic learning curve example concerns computing power. Moore's Law holds that computing power will double about every two years for the same cost. Sixty years after Gordon Moore, Intel's founder and former CEO, made this observation, the law holds true, though we're actually doubling now about every year. Kurzweil discovered, however, that Moore's Law is just the latest paradigm keeping a much longer trend of increasing computing power going, as Figure 2 shows.

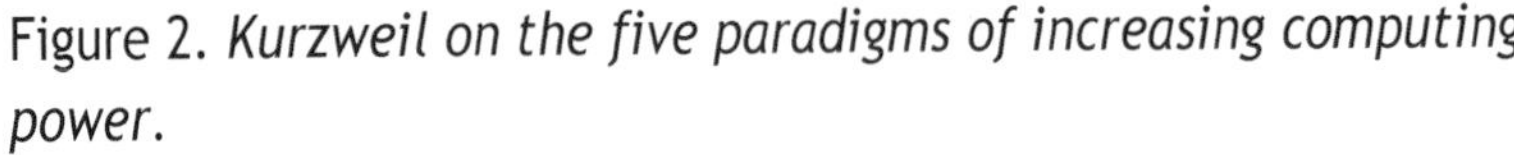

Figure 2. *Kurzweil on the five paradigms of increasing computing power.*

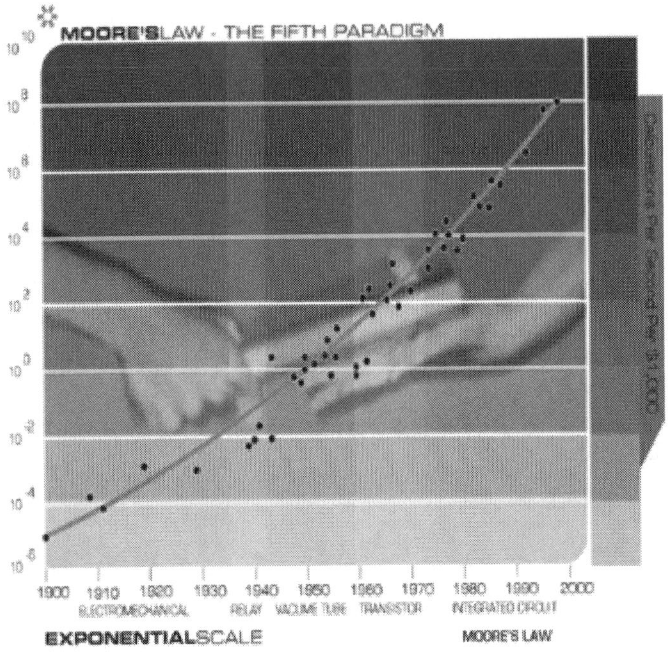

So how can 1% of all global electricity sales suggest in any way that we're on the road to a world dominated by solar? Here's why: 1% is halfway between nothing and 100%--in terms of doublings of capacity. That is, there are seven doublings from 0.01% to over 1% and also seven doublings from 1% to 100%. One double is two, which doubled is four, etc., reaching 128 after seven doublings.

Kurzweil's best example of the counterintuitive nature of his law is the Human Genome Project. This was a government-funded effort to decode the entire human genome in about fifteen years. Well, halfway through the project the effort was only at about 1% completion. Many observers wrote off the effort as a failure that couldn't possibly reach its goal. Kurzweil, however, wrote at the time that the game was won because 1% was halfway to 100% in the terms that actually mattered: the rate of improvement in sequencing technology.

And he was right. The Human Genome Project finished slightly early and helped to bring down the costs of genome sequencing technologies by orders of magnitude, as well as dramatic reductions in the time it takes for sequencing. We can now pay less than $1,000 to sequence our own genome in a matter of days.

Applying the Law of Accelerating Returns to solar

The average global growth rate for solar from 2008-2013 was 34%. NPD Solarbuzz projects 50 GW installed in 2014, for a growth rate of 27%, giving a six-year average of 33%. IEA calculates that solar PV will produce 160 terawatt hours in 2014, or 0.85% of global demand. Seven doublings of 0.85% gets us to 108%. So what is the doubling time at 33%? There's a simple cheat formula: 72/growth rate = the rough doubling time. So 33% leads to a doubling rate of a little more than two years (2.2 years).

Seven doublings of 2.2 years equals 15.4 years, at which point we would reach a solar penetration of over 100% of current global electricity demand by 2029 *if recent growth rates continue*. (Of course, global electricity consumption is going to increase in that time, but for simplicity's sake I'll keep my analysis to today's consumption.)

Will we actually get100% of our electricity from solar in just 15 years? Of course not. The growth rate will surely slow, as it has already in the last couple of years. This is to be expected: new technologies often see rapid growth rates as they catch on and then the growth rate slows as higher penetrations are reached. If solar continues to grow in coming years at an average rate of "only" 15%, we reach about 8% of current global electricity demand by 2030.

This is good but not great, considering the magnitude of the problems we face in terms of our fossil fuel energy use.

So are we going to see a declining rate of growth, steady growth or perhaps even an increasing rate of growth as solar becomes even more affordable in more and more countries? Only about a dozen countries were responsible for the vast majority of solar PV growth in 2013. Will this revolution spread quickly to other countries or will the solar revolution catch on fire around the world? So far, it's been a European, East Asian and North American phenomenon. Will the rest of Asia, Africa and South America catch up soon?

My expectations are certainly on solar catching fire (in the good sense) around the world now that it has become so cheap. We can expect Europe's transition to renewables to find renewed vigor now that Russia has shown increased aggressiveness in its backyard since early 2014. While the center of gravity in PV installations in 2013 swung sharply toward Asia (China and Japan were by far the largest two markets in 2013 and 2014), my feeling is that we may see it swing back toward Europe in 2016 and later as policymakers in Europe see the light on increasing their dependence on renewables as an alternative to continued dependence on Russian fossil fuels.

Time will tell what the rate of growth is in future years, but we can rest assured that solar will play an increasingly strong role in power grids around the world now that costs are so low, and steadily decreasing, and the stigma against solar is all but gone. It also seems fairly clear that solar power and other renewables are on a path to becoming by far the most prevalent form of power in most countries in the next few decades.

Since I wrote the first version of this chapter the IEA produced some major support for my contention that solar can become the dominant power source in the coming decades. Their November

2014 World Energy Outlook states that renewables—a combination of hydropower, wind and solar—are expected to exceed coal generation by 2040.

Even better, IEA stated in a pair of reports released in September of 2014 that solar could become the world's biggest electricity source by 2050, at 27% of global supplies. Solar is not exactly a toy anymore...

What about the U.S.?

We can't forget our own backyard in this discussion. Major investment banks are finally starting to see the potential for solar in the U.S., including Morgan Stanley and Citigroup. The U.S. became the third largest installer of PV in 2013, behind only China and Japan, and continued that pace in 2014, with about 6.2 gigawatts installed, a new U.S. record and a 30% growth rate over 2013. California, as usual, led the pack with over half of that growth (over 3.5 gigawatts). The U.S. also regained the top spot in Ernst & Young's Renewable Energy Country Attractiveness Index in 2013.

Figure 3. *U.S. solar growth through 2014 (source: GTM Research and SEIA).*

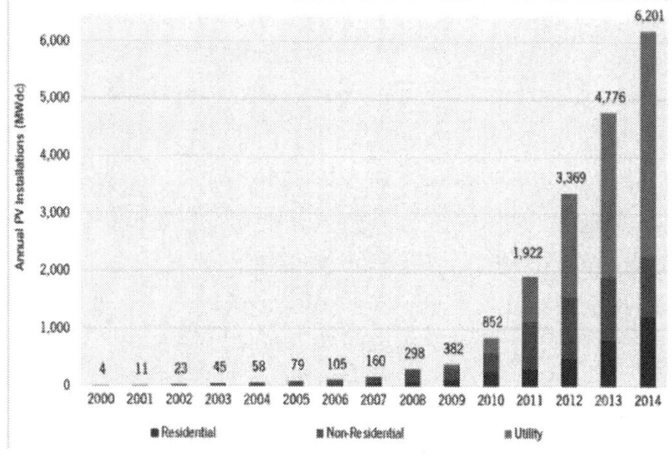

As solar reaches higher penetration levels in various countries and grids, integration issues will become more pronounced. I live part-time in Hawaii, on the Big Island, where solar penetration was already over 10% in 2014—one of the highest penetrations in the world. This is because electricity costs are sky high in Hawaii. High penetration of renewables can be managed in various ways but energy storage solutions are the most promising long-term solution. We are already seeing jurisdictions like Hawaii, California, Germany and Japan investing heavily in energy storage so it is reasonable to believe that we'll see a declining cost curve for storage technologies similar to what we've seen with solar. The "x factor" regarding storage is when this declining cost curve will take effect—in the next few years or further out? As we've seen with solar in Chapter 3, the learning curve phenomenon does allow for temporary plateaus or even increases in prices due to increasing commodity prices or lagging production. It is possible that global energy storage demand will exceed production and if that is the case we may see prices increase. But we can also expect that markets will quickly respond by increasing production and the long-term trend will be downward pricing.

Summing up, under Kurzweil's Law of Accelerating Returns, we have reached, after over forty years of development of solar PV technologies, the halfway point to market dominance, in terms of doublings of installed capacity. We are likely to see the second half of this journey take less time than the first half. Cost will probably not be the hurdle as we move forward because we are already at the point of cost-effectiveness in many circumstances, and on the verge in other circumstances, for solar in comparison to alternatives. Rather, integration issues and the stranded costs of existing energy infrastructure will be the larger hurdles.

Far-sighted businesses and policymakers should recognize the potential for solar and energy storage to provide a complete power solution in the coming decades, and be very careful in approving new long-term contracts for fossil fuel power production, because ratepayers may very well be on the hook for stranded costs for these new facilities as the solar revolution continues to work its magic.

Chapter 8

Crowdsourcing energy: how to transform an energy grid "overnight"

Crowdsourcing is taking off in almost every area one can imagine, including energy. In fact, crowdsourcing energy was one of the first areas to be crowdsourced, but under the very unsexy name of "feed-in tariff." Congress passed the first crowdsourcing of energy law, under an also-very-unsexy name: the Public Utilities Regulatory Policy Act (PURPA), in 1978.

PURPA required states to create programs under which utilities offered a set price to anyone who could produce power from a "qualifying facility," which included renewable energy like wind or solar power. The price was made public beforehand and long-term contracts were offered. These features created the market certainty that, lo and behold, created the first boom in renewable energy in the world, right here in California, in the 1980s and early 1990s.

PURPA was responsible for an "embarrassment of riches," as the California Public Utilities Commission described it at the time, in terms of a boom in renewable energy projects. This era created the global wind industry, centered in California for the first decade or so of global development. From those modest beginnings in Altamont Pass, Tehachapi and the San Gorgonia areas of California, the global wind industry will easily surpass 400 gigawatts of installed capacity in 2015. This is enough power for about 120 million California-size homes. That's real scale, made possible through a policy that essentially crowd-sourced power production in California in the 1980s under PURPA.

A couple of decades later, Germany's similar policy did the same thing for solar power. We're now at about 200 gigawatts of solar power installed worldwide, enough for about 40 million California-size homes (solar power's capacity factor is lower than wind, about 20% and 30%, respectively). Solar is growing much faster than wind nowadays, so solar will probably surpass wind in a few years. 2014 was the first year that solar and wind power capacity additions were approximately equal, at about 50 gigawatts each. We can expect solar power to start catching up with wind power quickly.

Germany's energy miracle—they have reached more than 30% renewable electricity in about 15 years—was made possible by policies that unleashed the ability of every Joe and Jane farmer to make money from selling power to the grid, as we saw in previous chapters. This is the essence of what a feed-in tariff is, the essence of what crowdsourcing energy is. The crowd is big in Germany, with over 1.1 million homeowners, landowners and businesses selling solar power from systems 10 kW and below at the end of 2013. A 10 kW system can fit on a small barn roof. Almost one million of these systems were installed from 2010-2013.

Germany has now installed almost 40 gigawatts of solar, in a bit more than a dozen years. This is enough solar to at moments supply up to 50% of German peak power demand, and about 7% of total system power. Only Italy gets more of its total electricity from solar, at almost 8% in 2014.

In fact, solar PV was the largest source of new generation in Europe in 2011 and 2012. This is remarkable for a technology that has been pooh-poohed for decades as an inconsequential plaything of environmentalists. A recent report from the International Energy

Agency found that 61% of solar installations in 2012 were driven by feed-in tariffs. There is now an increasingly large global crowd that has been literally empowered as mini-utilities by feed-in tariffs.

Another very impressive recent example of the ability of crowds to produce substantial amounts of energy even more quickly can be found in Japan. After the Fukushima nuclear disaster in early 2011, which led to the entire nation's fleet of nuclear plants being shut down, Japan created its first broad-scale feed-in tariff. This new policy produced almost seven gigawatts of solar power in 2013 and was expected at the time of this writing to reach ten gigawatts installed in 2014. Seven gigawatts of solar is as much power as a large nuclear plant, adjusting for capacity factor (a large nuclear plant is about two gigawatts capacity but a much higher capacity factor than solar). It takes at least a decade to build a new nuclear plant, and this was achieved with solar power in just one year in 2013, and then again (and then some) in 2014.

Figure 1. *Comparing solar installations through 2014 in different jurisdictions (source: various).*

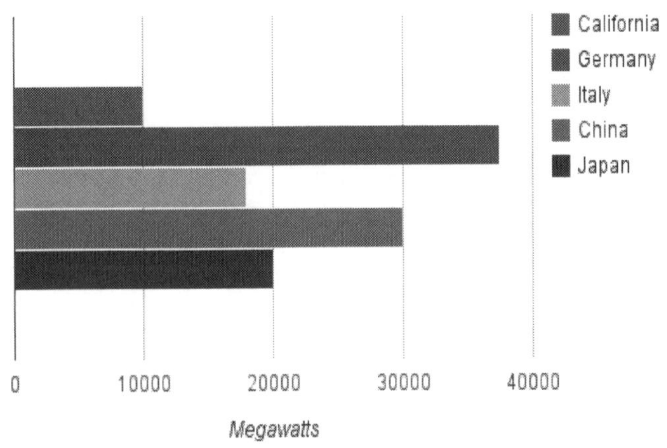

What about costs?

Just as important, these policies have helped dramatically reduce prices for these technologies. It's a virtuous cycle: as feed-in tariffs help deploy solar and other renewables at scale, the prices are reduced, and the reduced prices make it possible to deploy even more solar at less cost. We've seen in previous chapters how this trend, called "Swanson's Law" in the solar context, is highly quantifiable. Countries that created large feed-in tariffs can be directly credited with substantial price reductions based on the scale achieved through these programs.

Some observers criticize the German example, however, as an expensive exercise in "renewables at any cost." The prices paid did indeed start at a fairly high level, but prices have come down dramatically in recent years - from 55 euro cents per kilowatt hour in 2004 for systems between 40 kW and one MW, to just 12 euro cents per kilowatt hour in 2013. That's a reduction of almost 80% in nine years. This is known as "degression" and the point, of course, is to incentivize major cost reductions over time through planned degressions.

We are now at the point where solar power can compete with fossil fuels, without subsidies, for peak power, and many jurisdictions are now seeing solar be competitive even with non-peak power, particularly as energy storage technologies undergo the same transformation that the wind and solar power technologies have experienced. Scale is the best way to reduce prices for any technology, so as storage technologies get deployed at scale it is all but certain that costs will come down dramatically for this new technology as it has for solar.

Wind power has also come down in cost dramatically. A recent study from Lawrence Berkeley Laboratory, which produces annual reports on the wind and solar markets, found that the cost of wind power fell 39% in low wind speed areas from 2008-2013. This is because turbines are getting bigger and, in particular, longer blades are being used in order to harvest more wind. The cost has fallen a bit less in higher wind speed areas. Wind power has been competitive with wholesale power rates for a number of years now, so cost is not an issue for wind power.

Chapter 9

Pulling the plug on oil: going electric on transportation

Electric vehicles have been all the rage in the last few years, but those who have studied EVs and their history know that this isn't the first go-round at the rodeo for this "new" technology.

In fact, EVs were the most popular type of vehicle at the dawn of the vehicle era. In 1899, more EVs were sold than gasoline-powered cars. Albeit, the numbers were small (about 1,500 EVs were sold in that year) but nonetheless EVs were the dominant technology for at least a decade. But Mr. Ford came along in 1908 with his Model T and, along with the boom in oil production in the U.S., ensured the ascendancy of petroleum-powered cars for the next century.

There were a number of fits and starts in the EV industry, every couple of decades, but at no point were EVs anywhere near as dominant as they were at the dawn of the transportation revolution in the last part of the 19th Century. Most adults living now are aware of GM's EV1 and the tawdry history of that vehicle in the 1990s. There was even a movie made about it: *Who Killed the Electric Car?* (And a followup in 2011).

It seems, however, that this latest phase in the development of EVs is likely to be more than a blip in automotive history. The new era for EVs began in late 2010 with the advent of the Chevy Volt plug-in hybrid car and the Nissan Leaf all-electric car. EVs have come on strong in the last four years, with dozens of EV designs either on the market today or soon to be. And sales are picking up quickly, albeit from a very modest starting point.

Norway is leading the world in terms of EV sales as percent of market share. In November of 2013, EV market share for new sales reached a staggering 12%. California has reached about 1.2% of all cars sold and comprises about one-third of U.S. EV sales. California broke through the 100,000 vehicles mark in mid-2014 and is well on its way to the 200,000 mark, though sales slowed down in late 2014.

In terms of the U.S. as a whole, we are fast approaching 1% of all sales coming from EVs (pure battery electrics, BEVs, and plug-in hybrid electrics, PHEVs, but excluding hybrid car sales). EVs comprised about 2/3 of a percent for all light duty car sales in the U.S. in 2013 and just a little higher (70%) in 2014—a disappointing stall in the growth rate of EV sales. We may hit 1% in 2015 as a number of new EV models become available.

Globally, we're at about the same level. Total EV sales in 2014 were 320,000 and cumulative sales reached 740,000 by the end of 2014. If sales in 2015 reach 500,000 or more, as they are likely to do, EVs will be about 0.6% of all passenger vehicles sales (about 89 million in 2015, according to HIS Global Insight). EV sales are likely to reach 1% of all sales in 2016. 1% is still very small—way too small—but it turns out that reaching 1% is a really important milestone. In fact, the game may be won when 1% is reached under very similar reasoning to what we saw in Chapter 8 in the discussion about exponential growth in solar.

Applying the Law of Accelerating Returns to EVs

It seems incredible but the following is a mathematically true statement: if customers continue to buy EVs at the same 100% rate of growth as we saw in 2012 and 2013, all cars sold in 2020 would be EVs. This is because seven years is seven doublings and seven doublings from 1% gets us past 100%. Of course, the "if" in my statement is the key. No one should reasonably expect that we'll see 100% rates of growth in EV sales every year through 2020—we won't. We already saw the growth rate slow to 80% for 2014 and even lower in 2015. Figure 2 shows the national rankings by share of the growing EV market in 2014, with the U.S. practically tied with China for #2 after Japan. Cumulatively, the U.S. beats China easily, with a 25% share of all-time EV sales compared to China's 11%, but still far behind Japan's impressive 43% share (Figure 3).

Figure 1. *National EV sales shares (data source: http://ev-sales.blogspot.com).*

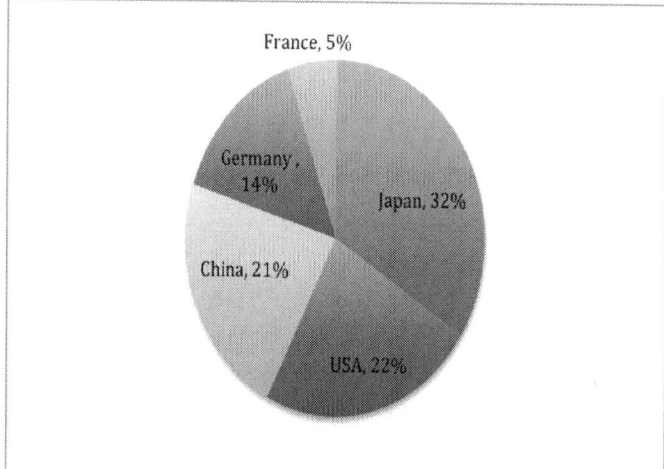

Figure 2. *Cumulative all-time sales of EVs by country share.*

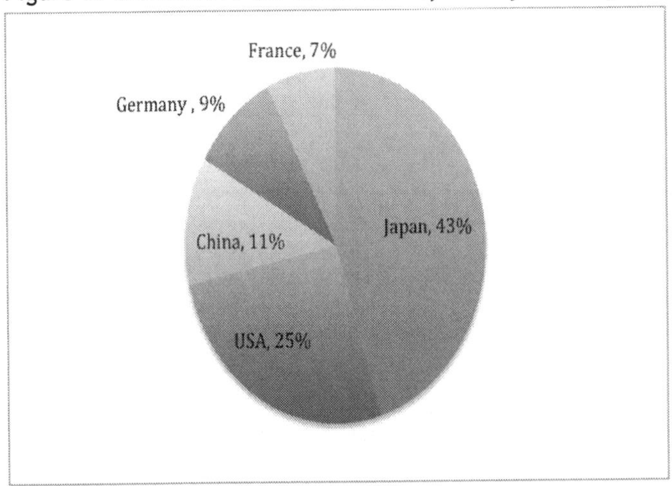

The rate of global growth will surely slow down substantially as various obstacles present themselves. U.S. sales growth stopped in 2015, but this is likely to be an anomalous year because of new versions of the Chevy Volt and the Nissan Leaf coming out. Even if the U.S. rate of growth averages “only” 50% each year, we would reach 100% of all sales by 2027. This won’t happen either, but it is reasonable to expect that a substantial percentage of all cars sold will be EVs by the mid-2020s. Given the number of new models coming on to the market by 2017 (there will be over 40 EV models available by then in the U.S.), including the Chevy Bolt and the Tesla Model 3, which are both projected to cost around $30,000 and have over 200 miles of range, plus in particular the rapidly declining costs of batteries, I think it’s reasonable to expect at least 30% average annual sales growth through the mid-2020s. The Bolt and Model 3 may well be the game changers that the industry has been waiting for. With projected 2017 availability we won’t have to wait very long to find out if these cars live up to expectations.

Figure 3 shows the result of various average growth rates over time. Even at 20% annual growth EVs comprise almost half of all cars sold by 2034.

Figure 3. *U.S. EV sales as a function of various growth rates.*

Year	50%	30%	20%	10%
2015	1%	1%	1%	1%
2016	2%	1%	1%	1%
2017	2%	2%	1%	1%
2018	3%	2%	2%	1%
2019	5%	3%	2%	1%
2020	8%	4%	2%	2%
2021	11%	5%	3%	2%
2022	17%	6%	4%	2%
2023	26%	8%	4%	2%
2024	38%	11%	5%	2%
2025	58%	14%	6%	3%
2026	86%	18%	7%	3%
2027	130%	23%	9%	3%
2028		30%	11%	3%
2029		39%	13%	4%
2030		51%	15%	4%
2031		67%	18%	5%
2032		87%	22%	5%
2033		112%	27%	6%
2034			32%	6%
2035			38%	7%
2036			46%	7%

Barriers to adoption

The Law of Accelerating Returns isn't really a law, of course. There's nothing inevitable about the development of particular technology or adoption curves. If there was the Betamax video player would be in every household today. While Kurzweil's data shows that many technologies do follow the traditional S-shaped development curve and continuously improve, the fact remains that the vast majority of new technologies and new ideas don't become ubiquitous.

So the hard question is: why do some technologies become ubiquitous and some not? And in the case of EVs, are we on the road to ubiquity or are there a number of roadblocks preventing a future in which every garage or street curb has an EV keeping it company?

There are a number of obvious obstacles right now that are keeping sales figures in the very low single digits in the U.S., including, among others: 1) lack of widespread awareness about the availability and benefits of EVs; 2) high upfront cost of EVs, driven primarily by battery costs; 3) range anxiety due to insufficiently widespread public charging stations and similar concerns about the speed or cost of charging; 4) tough competition from other types of increasingly fuel-efficient vehicles.

I'm going to focus in the rest of this essay on what is probably the most difficult obstacle to overcome, battery costs, because the solution is not simply a matter of throwing money at the problem.

Kurzweil's Law is a technology improvement law, rather than a technology adoption law, so it's not directly applicable to adoption rates for solar or EVs, even though I've treated it as such above (my bad). However, it's *indirectly* applicable because the adoption rate of EVs is highly dependent on technology improvement; specifically, the rate of improvement of the most expensive (by far) component of EVs—the battery.

Battery costs are already falling rapidly with increased deployment of EVs and other uses for battery technologies. A 2013 report from McKinsey & Company projected "dramatically" falling prices for batteries by 2020—to around $200 per kilowatt hour, and $160 by 2025, down from $500-600 at the time the report was written in

2012, and down from about $1,200 in 2009. A 2013 report from Navigant Consulting agrees in general with the McKinsey team, projecting $180 by 2020, down from $500 in 2014. A 2015 analysis of Tesla's new stationary battery business concluded that Tesla was already selling their new batteries at about $250 per kilowatt hour, so we may be seeing battery price declines even faster than expected.

Navigant also projects a more than ten-fold increase in EV battery production by 2020. If we see the same price drop in this area that we have seen with respect to solar panels—a 20% drop with every doubling of production—we can expect a bit more than three 20% price drops before 2020. This amounts to a net cost reduction of about 55%, which is less than the drops projected by Navigant and McKinsey, but definitely in the same ballpark. This quick calculation, plus 2015 price data, gives a little more confidence that these companies' projections aren't outlandish.

A 2015 study found that battery costs had declined an average of 8% since 2007 and Tesla's battery costs had declined an encouraging 14% per year in the same time frame. The same study found that Tesla's vehicle battery costs were already down to $300 per kilowatt our in 2014.

These data, combined with projected battery price reductions, would bring the cost premium of an EV to only about $2,000 by 2020, when compared to a comparable conventional auto. This premium can quickly be earned back with fuel savings, due to the cost of electricity being so much lower than the cost of gasoline (about 1/3 to 1/2 the price generally), so the total cost of ownership of EVs will at that point be far less than for conventional vehicles.

In sum, while we may not be on the road to an inevitable "EV in every garage" future just yet, due to a number of persistent obstacles, it seems that we are indeed on the cusp of such inevitability. "Just" seven more doublings and all cars sold will be EVs.

I feel less confident about EVs becoming ubiquitous than I do about solar becoming ubiquitous, largely because EVs are a much newer technology and thus have less of a track record, but this analysis shows that the future could well be an EV future under what are reasonable cost reduction scenarios. The real game changers in this space will be the success or failure of the Chevy Bolt, the Tesla Model 3, and similar EVs that are slated to provide 200+ miles of range for a price tag of $30,000 or so. Both of these companies are planning launch dates of around 2017, so it won't be long before we get a better idea of where the EV market is headed in the next decade.

There are already reports of many possible imitators of these two market leaders and the-more-the-merrier philosophy holds here because the more competition there is for affordable EVs with good range the quicker the public will be able to realize the benefits of low-cost electric vehicles.

Chapter 10

My personal EV solution

I drove a Prius C for two years and it just felt too big. Even this itty-bitty car dragged around about 2,500 pounds of steel and plastic while it also dragged me around. And it felt like a mini-tragedy each time the gas engine kicked in when the car reached about 15 miles an hour. No more pure electric drive, I was burning squashed dinosaurs again.

Don't get me wrong, the Prius C is a great little car and when I got it in mid-2012 it was the most fuel-efficient vehicle you could buy. For a while now I've been increasingly interested in the energy efficiency of transportation. Today's cars, even tiny ones, are extremely inefficient in terms of how much energy is wasted in getting generally just one person around. Even the Prius C wastes a good chunk of the energy content of the gasoline fuel it uses, due to inherent inefficiencies in gasoline engines. Generally speaking, internal combustion engine light-duty vehicles (passenger cars and light trucks) lose almost 60-75 percent of the gasoline's energy through heat losses. The rest is lost through idling, braking losses, idling losses, and other factors.

We can clearly do a lot better than wasting 85 percent of the fuel we use in transportation.

I personally pulled the plug on oil by diving fully into the world of electric vehicles in late 2014. Fiat offered a sweet deal on a lease for the Fiat 500e in 2014. This little car is not only highly efficient;

it's also super stylish, even sexy in a geek chic kind of way. Here's a portrait of my new love (Figure 1).

Figure 1. *My new Fiat 500e smiling for the camera.*

The Fiat was available under a 39-month lease at $199/month and about $1,000 down. This amazing deal was possible primarily due to three things: 1) California's fairly aggressive Zero Emission Vehicle mandate, which requires an increasing number of zero or low emission vehicles to be sold by each brand; 2) a federal $7,500 tax credit (not a deduction, a credit) for all new pure electric vehicles sold; 3) rock-bottom interest rates.

The kicker is that California also offered a $2,500 rebate for pure-electrics so when this payment was factored in the Fiat was almost free for the first year of its ownership. Yes, free.

Deciding to take the plunge

I'd thought about going pure-electric for a while but hadn't until until I bought my Fiat taken the plunge, for a few reasons, including the fact that I'm a single guy in a one-car household and having an

EV with limited range wouldn't work for my lifestyle. Cost was also a concern.

The Fiat 500e has an official range of 87 miles. After my first six months of ownership, that range is a bit optimistic. My average range is about 75-80 miles per charge. I usually live in Santa Barbara and make occasional trips to Los Angeles and San Francisco, and I also go on longer road trips camping or visiting family in Washington State and Oregon a couple of times a year. Clearly, those road trips won't be possible in a car with 80 or so miles of range. But here's the cool thing about the Fiat lease: it comes with $500 (about 12 days) of free car rental per year, and that's what gave me the confidence that an EV could work for me.

There are a number of EVs available today in California and other states (the Fiat is only available in California and Oregon for now) and I did a fair amount of research before settling on the Fiat. The primary models available in 2014 included:

- The Nissan Leaf, which is by all accounts a great car but its aesthetics are wanting and my feeling is that green technologies should and can be sexy
- The Tesla Model S. This is, like about every guy my age, my dream car. But the closest I can come to owning it right now is drooling on ones I see in parking lots around town. Its base price is $72,000 and it can easily reach about $100k when various options are added, like dual drive and autopilot. One day I'll own a Tesla, but not now. If I took on that above-$1,200 monthly lease it would own me, not I it.
- The Chevy Volt plug-in hybrid. Owners love this car and a new survey found that 92 percent of owners are very happy with their electric vehicle,

The Chevy Volt would have been a better fit for me than the Fiat because the Volt has a gas tank that can be used to charge the batteries for much longer range than the 38 miles of the built-in batteries. However, my situation was unusual in that my car loan on my Prius C was upside down by about $5,000. Fiat gave me a pretty good deal on my trade-in so even with the upside down portion rolled in to my new lease it's still affordable. My local Chevy dealer couldn't offer me a very good deal, so despite my appreciation for the Volt, and its greater practicality in terms of much greater range and more passenger and storage space, I opted for the Fiat.

I don't regret that choice at all. The Fiat is a really fun car to drive. It's very zippy and you can make the wheels chirp without much effort. It's so cute you'd think it's a Hello Kitty car rather than a Fiat. And it's Italian-made, with all that that entails. Its various features work wonderfully and it's well-made. It even has heated seats. The zinger for my particular model is the sunroof. I highly recommend the sunroof on this vehicle since it can be a vent, an open sunroof, a skylight, or a screened sunroof, depending on your needs at the moment.

I was somewhat amused when the dealer told me that no service would be needed on my new car until 20,000 miles, but with the version with a sunroof it would be 15,000 miles. This is another major advantage of EVs: far fewer moving parts, which don't require anything like the servicing that ICEs need.

EVs can be a great deal now due to the subsidies offered by federal and state governments. But we can already see the future clearly: battery costs will continue to come down dramatically, vehicles will become steadily lighter, and we will as a consequence see EVs becoming increasingly affordable even without subsidies. Moreover, the current federal and state subsidies are designed to phase out over time volumetrically, so today's subsidies aren't designed to be perpetual. This phaseout will act as an incentive to manufactures to work hand on reducing prices.

EVs are inherently more efficient than internal combustion engines

Now, back to the efficiency discussion with which I began this piece. In my case, I charge at home (I work out of my home), or at often free public chargers in town, and electricity is part of my rent so I now pay about zero for fuel costs. But even if I paid for electricity I'd still see major fuel cost savings because EVs are inherently far more efficient than ICEs. As mentioned above, ICEs waste about 85 percent of the energy they use. EVs waste only about 40 percent of power from the grid, and even less if that power is received from solar panels on your roof.

A good way to compare ICEs and EVs is the "miles per gallon of gasoline (mpgge)" measure, which is designed to convert electricity as a vehicle fuel into gasoline equivalents. For example, my Fiat has a 117 mpgge rating, compared to 50 mpg for the 2014 Prius C and 35 mpg for the Nissan Versa. See Figure 2 for the full comparison. So comparing equivalent small cars to my Fiat we get a fuel efficiency bost of 2-3 times. And if we compare larger cars we get an even more favorable result.

Figure 2. *Comparison of Nissa Versa, Toyota Prius C and Fiat 500e (source: www.fueleconomy.gov).*

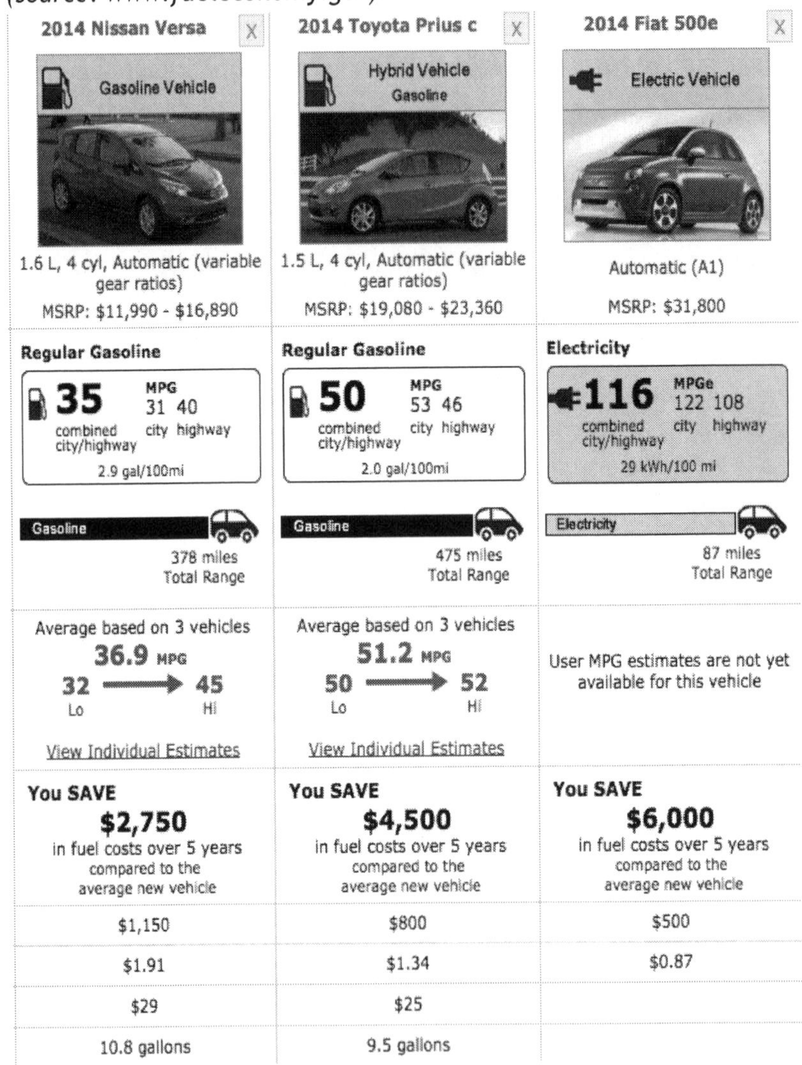

2014 Nissan Versa	2014 Toyota Prius c	2014 Fiat 500e
Gasoline Vehicle	Hybrid Vehicle Gasoline	Electric Vehicle
1.6 L, 4 cyl, Automatic (variable gear ratios)	1.5 L, 4 cyl, Automatic (variable gear ratios)	Automatic (A1)
MSRP: $11,990 - $16,890	MSRP: $19,080 - $23,360	MSRP: $31,800
Regular Gasoline	Regular Gasoline	Electricity
35 MPG combined city/highway; 31 city, 40 highway; 2.9 gal/100mi	50 MPG combined city/highway; 53 city, 46 highway; 2.0 gal/100mi	116 MPGe combined city/highway; 122 city, 108 highway; 29 kWh/100 mi
Gasoline 378 miles Total Range	Gasoline 475 miles Total Range	Electricity 87 miles Total Range
Average based on 3 vehicles 36.9 MPG; 32 Lo → 45 Hi; View Individual Estimates	Average based on 3 vehicles 51.2 MPG; 50 Lo → 52 Hi; View Individual Estimates	User MPG estimates are not yet available for this vehicle
You SAVE $2,750 in fuel costs over 5 years compared to the average new vehicle	You SAVE $4,500 in fuel costs over 5 years compared to the average new vehicle	You SAVE $6,000 in fuel costs over 5 years compared to the average new vehicle
$1,150	$800	$500
$1.91	$1.34	$0.87
$29	$25	
10.8 gallons	9.5 gallons	

Based on this inherent increase in efficiency of EVs, fuel costs are one-half to one-third as much as fuel costs for equivalent ICE cars. As Figure 2 shows, the fuel cost savings of my Fiat are about $6,000 over five years when compared to the average new vehicle.

For the EV revolution to continue we'll need to see battery costs steadily fall in price. Currently they still add up to a significant premium for EVs compared to their non-electric cousins. For example, the sticker price of my Fiat 500e is about $31,000, compared to about $17,000 for the non-electric version. That's a premium of $14,000, which isn't currently made up for with fuel-cost savings. But if that premium falls by half in the next few years, which is a reasonable expectation, we may well see many EVs become affordable on their own terms when we consider the fuel-cost savings. At the same time, many premium features come with switching to EVs.

Let the revolution roll on.

Chapter 11

California steps up again on EVs

Say what you will about California but it's undeniable that we've been leaders when it comes to transportation and vehicle policy. When it comes to electric vehicles (EVs), in particular, California is showing the rest of the country the art of the possible. Sales of EVs in California exceeded 100,000 in 2014 and we're well on our way to doubling that figure.

The center of the action on EVs is, perhaps surprisingly, the California Public Utilities Commission (CPUC). I say "surprisingly" because normally on this kind of issue the Legislature in Sacramento would be calling the shots. In this case, however, lawmakers have given a large amount of discretion to the CPUC (as they also did with energy storage legislation) to determine its preferred policy.

The regulated utilities—the big three of Pacific Gas & Electric, Southern California Edison and San Diego Gas & Electric—have also been stepping up recently with their own proposals for kickstarting the EV revolution.

SB 626, passed in 2009, directed the CPUC to "evaluate policies to develop infrastructure sufficient to overcome any barriers to the widespread deployment and use of plug-in hybrid and electric vehicles ..." And that's about all the law directed. The CPUC did act on this law by the required July 2011 deadline by issuing the first in a string of major decisions (this is how the CPUC promulgates regulations).

The Governor also got into the game in a big way in 2012 by issuing an Executive Order that set forth a number of mandates for the CPUC and other state agencies to implement. Even though the CPUC is an independent agency, Governor Brown still believes that he can set policy and he set many major policies in his 2012 order, including requiring that by 2015 all major cities be EV-ready, by 2020 California has sufficient charging infrastructure for up to one million EVs and by 2025 at least 1.5 million EVs and fuel cell vehicles (collectively "zero emission vehicles" or ZEVs) be on the road.

The big news recently, however, comes from the state's three big private utilities, San Diego Gas & Electric (SDG&E), Pacific Gas & Electric (PG&E) and Southern California Edison (SCE). SDG&E submitted an application in early 2014 to the CPUC to own 5,500 public charging stations. SCE one-upped SDG&E in a big way with an application to the CPUC to develop the infrastructure for up to 30,000 EV charging stations in the next few years. SCE will develop the required distribution lines, transformers, and other infrastructure and also provide rebates of up to $3,900 for third parties to own the EV chargers. This is known as a "make ready" approach because it relies on the utility making the infrastructure ready for third parties to build out the charging stations themselves.

The price tag for SCE's proposed program is about $350 million, to be paid for with ratepayer funds. This is not chump change but SCE asserts, without much demonstration yet (it's early in the process of regulatory approval so this is not unusual), that its program will be cost-effective for ratepayers. While we need a lot more evidence to firm up this conclusion my feeling is that we know enough from many similar analyses to feel pretty confident that SCE is right on this. The key savings from EVs come from the fuel savings, simply because running cars on electricity rather than petroleum is so much more efficient (EVs are 2.5-3 times more efficient than gasoline vehicles in terms of energy used).

Investments in EV infrastructure are a highly cost-effective proposition for ratepayers and taxpayers because EV transportation is so much more efficient and thus cheaper than fossil-fuel based transportation. The Transportation Electrification Assessment, a report from ICF International that SCE cites in its application, calculates the costs and benefits for various scenarios, as summarized in Figure 1.

Figure 1. *ICF Int'l conclusions regarding cost-effectiveness of EVs.*

PEV	Private B-C Ratio	Societal B-C Ratio	Total
PHEV10 - PC	12.53	1.54	14.07
PHEV10 - LT	7.80	0.97	8.77
PHEV20 - PC	7.49	1.13	8.62
PHEV20 - LT	4.48	0.65	5.13
PHEV40 - PC	3.84	0.67	4.52
PHEV40 - LT	2.96	0.48	3.44
BEV - PC	8.89	1.28	10.17
BEV - LT	3.86	0.55	4.41

A "B-C Ratio" over one means that there are positive net benefits. Every scenario examined has a benefit-cost ratio over three, which means benefits outweigh the quantified costs by at least a factor of three for all vehicle types examined. This is a very encouraging result in terms of the ability of EVs to produce net economic benefits for ratepayers and taxpayers.

SCE is asking the CPUC to approve their major new program in two phases: Phase 1 will be a $22 million one-year pilot to test the concept. Phase 2 will be over $300 million and will take four years to complete. The point of SCE's program is to help California achieve the Governor's 1.5 million ZEV mandate by 2025. This is a big lift, even with California already at well over 100,000 EVs on the road today. We'll need a lot more infrastructure to instill confidence in mainstream consumers that they will be able to charge their EVs when and where they want to.

SCE's focus is on "long dwell-time" locations like apartments and workplaces. This is the case because SCE sees an opportunity to install numerous Level 1 and 2 chargers at these locations that will allow EV owners to charge up while they're working or doing whatever they do at home. DC Fast Chargers aren't included in SCE's proposal because SCE feels that the market for these types of chargers is growing adequately.

PG&E, not to be outdone, submitted its own application in early 2015 asking the CPUC to approve a program that would allow it to own over 25,000 charging stations. PG&E, somewhat surprisingly, is seeking to own the chargers rather than emulate SCE's "make ready" approach in which it would own the infrastructure required for the chargers but not the chargers themselves. Given the very negative response so far to PG&E's application it is all but certain that the CPUC will deny PG&E's request for direct ownership of that many chargers.

Business opportunities in the EV charger space

These major new programs provide significant opportunities for third parties to create viable business models. SCE's proposed rebate per charger of up to $3,900 means that third parties seeking to own EV charging stations in long dwell-time locations will probably pay nothing, or very little, for these stations. If this is the case, it means that the major costs will be the property lease or property purchase cost and the cost of power bought from the IOU for re-sale to EV owners.

The revenue options for long dwell-time charging locations are not as diverse as for charging stations focused on shorter dwell-times, which will generally include DC Fast Chargers (which can charge some vehicles 80 percent in just 20 minutes, as opposed to four hours or so for Level 2 chargers and far longer for Level 1 chargers). For long dwell-time locations, charging stations located in workplaces or multiple unit dwellings (MUDs) may offer charging for free as a perk of working or living at that location. Alternatively, monthly subscriptions could be offered for unlimited charging.

The new IOU programs may allow some DC Fast Chargers to be included at the long dwell-time locations, in which case there could be higher revenue potential due to faster turnaround. For example, PG&E's program does include 100 locations with DC Fast Chargers.

What does the future hold?

SCE's application makes the case for its proposed ratepayer funding based on the Governor's mandates for EVs, including the 2025 mandate of at least 1.5 million ZEVs on the road. Given the binding nature of SB 626 and the Governor's ZEV executive order, plus the fact that promoting EV adoption through better EV charging infrastructure is a clear economic winner for ratepayers and taxpayers, it seems likely that the CPUC will approve the IOU applications for EVSE ownership, at least in some form.

If this does happen, we'll have a much better chance of meeting the 2025 mandate of 1.5 million ZEVs. Since I recently took the plunge and bought an all-electric Fiat 500e, I can now speak from personal experience about the features and benefits of EVs. I think the combination of excellent driving features, amazing fuel economy (my Fiat gets the equivalent of 115 mpg), and low cost of driving will catch on widely in the coming years, particularly as battery costs continue to fall rapidly with scale.

My feeling, which isn't much more than intuition at this point, is that once we get to the scale of 1.5 million or so ZEVs on the road in California (the vast majority of which will be EVs rather than fuel cell vehicles), we'll have reached critical mass and EVs may well become the default purchase option for many new buyers, including fleets and individual buyers. If this is the case, California's long-term climate mitigation goals of an 80 percent reduction in greenhouse gas emissions by 2050 will become eminently achievable.

Chapter 12

Why hydrogen fuel cell vehicles don't make sense

The first mass-market consumer fuel-cell vehicle in the U.S. became available in late 2014: the Hyundai Tucson Fuel Cell SUV. It's taken decades to get to this point, so many enthusiasts are hoping that this will be a tipping point for the technology. It's an exciting development, and by all accounts thus far, it's a great car with a decent range of 250 to 300 miles. However, it's not at all clear that fuel cell vehicles make much sense from a policy perspective. This article delves into the details and compares fuel cell vehicles (FCVs) to electric vehicles (EVs) like the Nissan Leaf and plug-in hybrid electric vehicles (PHEVs) like the Chevy Volt.

For those who are skimming for quick insights, here's my main conclusion: FCVs don't make sense because the energy losses from creating hydrogen fuel and converting that fuel back into electricity are far too large. In fact, these losses mean that battery electric vehicles like the Leaf, Tesla or Volt, can drive 2-3 times as far on the same amount of electricity as an FCV. That's a very high, and probably an insurmountable, hurdle for FCVs to get over.

I've followed developments in this field for over a decade. I was the primary author of a blueprint for weaning Santa Barbara County from fossil fuels when I worked for the Community Environmental Council as its energy program director.

Here's an excerpt from that 2007 document: "Hydrogen vehicles -- fuel cell or hydrogen internal combustion engines -- hold some long-term promise. Toyota and General Motors have announced plans to offer retail vehicles in 2010 or soon thereafter, but most analysts believe it will be some time later that these vehicles are available to consumers at affordable prices."

We followed up on this short analysis a year later with our Transportation Energy Blueprint for Santa Barbara County, a more detailed look at the most promising solutions for fossil-fuel reduction in the transportation sector. Michael Chiacos, who is still with the Community Environmental Council, was the primary author of that report. We conducted another extensive public process as part of creating this document, and we ranked the various transportation solutions based on a number of quantitative criteria. FCVs didn't make the list because even though our team thought they had promise, the likelihood of that promise panning out was low and the potential impact of FCVs by 2020—the analytical timeframe of our second report—was, even under the best case scenario, very low.

I have since become even more negative when it comes to FCVs, and my worries about the ability of manufacturers to deliver on the promise of FCVs are starting to look pretty accurate. We are now in 2015 and there are only three commercially available FCVs on the market or soon to be on the market (Hyundai, Toyota and Honda). None of these companies expects much in the way of sales in 2014 or 2015, largely because of the lack of both a public fueling infrastructure and a home fueling option.

The lack of a fueling infrastructure really is a chicken-and-egg problem: we would need more FCVs to justify more fueling stations and vice versa. *Car and Driver* magazine stated in a recent article on the Hyundai Tucson, the first FCV to become commercially available, that "operators of the Tucson...will be confined to their respective metro areas." The article adds that home fueling stations are not "being considered at the moment." Though it is not explicitly stated in the article, it seems very likely that this lack of a home fueling option is due to safety issues.

The Tucson costs $499 per month, which includes free fueling (yes, free), with an additional $2,999 due at signing. So this car will not be cheap, but is certainly a lot cheaper than, say, a Tesla, which costs at least twice that much for a lease. The Tesla lease also includes always-free fueling at Tesla's Superstations around the country.

Hyundai plans to build up to 1,000 Tucsons each year if leases are snapped up. However, the company is projecting only about 300 sales per year at this point.

In terms of the fuel-cell infrastructure in California, there are only nine public stations currently, but 45 more are planned, with a major new round of funding recently announced by the California Energy Commission.

A more detailed comparison of FCVs and EVs

Despite the big head start that EVs have on FCVs, are there other reasons to support FCVs as a viable or even better alternative? There are some benefits associated with FCVs, which will be detailed in the following sections. But the alternative vehicle market has already experienced a number of false starts (methanol in the 1980s and ethanol in the last decade, to name just two), and it would be smart to avoid yet another false start by over-hyping the ability of FCVs to reduce fossil fuel use and greenhouse gas emissions in any serious manner. Let's examine the evidence.

Vehicle range

Vehicle range for FCVs is generally advertised as 200 to 300 miles, which is comparable to the distance many internal combustion vehicles can travel on a tank of gasoline. The Tucson, for example, is supposed to have up to a 300-mile range.

EVs generally have a much shorter range, with the Nissan Leaf, for example, having a real-world range of only about 80 miles per charge. The Tesla Model S, a far more expensive vehicle, has up to 300 miles of range, but the Tesla Model S is far from being a mass-market vehicle.

Plug-in hybrids (PHEVs) like the Chevy Volt actually have better range than most internal combustion vehicles or FCVs, with the 2014 model achieving real-world range of 350 to 400 miles with a tank of gas and a charged battery.

In this head-to-head matchup, PHEVs win, but FCVs come out ahead of EVs.

Refueling time

Refueling time is generally much shorter for FCVs than for EVs and PHEVs. It takes just a few minutes to refuel an FCV versus up to many hours for EVs and an hour or so for most PHEVs. As discussed below, Tesla is planning to offer a battery swap option, which makes refueling faster than gasoline or hydrogen refueling, but even if this does come to fruition, it probably won't be an option for most EVs anytime in the near future.

This category gives a clear win to FCVs.

Lifecycle fuel efficiency

Perhaps the biggest downside for FCVs arises from how the hydrogen fuel is created. From an environmental perspective, hydrogen would ideally be created through electrolysis of water, using renewable electricity to split water into hydrogen and oxygen, rather than from fossil fuels like natural gas. However, a lot of energy is lost in the FCV fuel cycle, through electrolysis of water and then conversion back to electricity in a fuel cell. In fact, according to a 2014 study from the University of California at Irvine,

EVs are 2.5 times more efficient than FCVs in using energy (Figure 1).

Figure 1. *UCI study comparing fuel efficiency of EVS, FCVs and other vehicle types.*

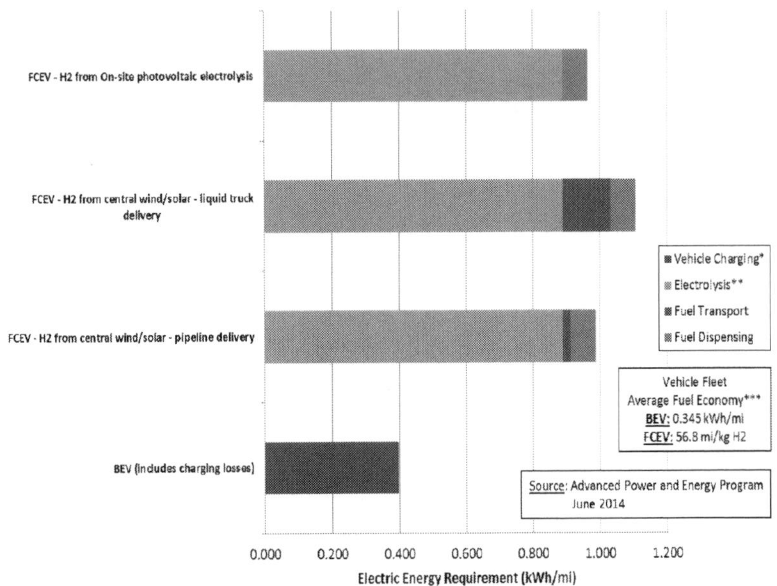

Based on this major difference in energy efficiency, it would be far more efficient to simply use the electricity, otherwise required to create hydrogen, to directly fuel an electric vehicle or a plug-in hybrid electric vehicle. This is a serious demerit for FCVs in a world of very limited energy.

Some FCV supporters have suggested that similar inefficiencies arise from EVs due to the line losses in charging from the grid. This is inaccurate for two reasons: first, the net line losses for most modern grids is quite low (California is only about 5-7%, for example); second, most EVs will charge in urban areas where much of the power consumed is produced locally, reducing line losses even further. And, increasingly, homeowners are charging from their own solar panels, with almost no line losses.

FCV advocates also argue that cheap renewables will make electrolysis sensible before too long. But if renewables are cheap enough for electrolysis to make sense, it would still make far more sense to use that electricity directly in EVs, with 2.5 times lower conversion losses. In sum, comparing net energy losses results in a strong win for EVs.

Fueling infrastructure and cost

As mentioned, California has only nine public hydrogen fueling stations but another 45 are currently planned. This is the situation a decade after the creation of the Hydrogen Highway Network in 2004, and this figure includes the 28 new stations recently announced by the California Energy Commission.

The key problem for adding new hydrogen infrastructure is the cost: charging stations cost about $2 million each (the CEC is awarding $46 million to construct 28 new stations, but this cost doesn't include matching funds), versus $5,000 to $15,000 for Level 2 EV charging stations and about $60,000 or more for DC fast chargers.

Adjusting for the number of vehicles that may be charged at hydrogen fueling stations (which is likely to be far greater than for EV charging stations due to the much-longer fueling time for EVs), FCVs still have higher infrastructure costs. This conclusion is bolstered if claims of future availability of battery-swapping stations from companies like Tesla are genuine. Battery swapping will, when available, allow refueling faster than gas or hydrogen fueling. Better Place pioneered this concept, but the company's efforts did not pan out, and it has since gone bankrupt. We'll see if Tesla's concept does better.

More importantly, all EVs can charge from a home outlet or workplace outlet if required, even though this takes ten to twenty hours for most vehicles. For faster home or workplace charging, many companies now offer Level 2 chargers that reduce the charging time by a factor of three or four.

The number of public EV charging stations has grown exponentially in the last decade, up from just a handful in 2004 to over 2000 public stations in California alone in 2014. Today, there are nearly 9,000 public EV charging stations in the U.S. We are also seeing an increase in the number of fast charging stations in California and elsewhere. Fast chargers are important because they allow an 80 percent charge in about 20 minutes for compatible vehicles, as opposed to a process that requires several hours with other types of chargers.

A final mark in favor of FCVs is that people without a garage or another place to charge an EV at home or at work won't be able to buy an EV. FCVs could make sense for these people. That said, there is a very large market of customers for EVs and PHEVs made up of people who do own their homes or otherwise have access to charging space.

EVs win in the fueling infrastructure category.

Vehicle availability

There are three FCVs planned for 2014 and 2015. In contrast, there are about two dozen EVs and PHEVs on the market in 2014 and another two dozen models that will become available in the next couple of years. However, there are only about sixteen models widely available in states like California, and sales are dominated by just five companies: Nissan (Leaf), GM (Volt), Tesla (Model S), Toyota (Prius Plug-in), and Ford (C-MAX and Fusion Energi). Cumulative sales in the U.S. exceeded 200,000 by the end of 2013, with sales figures tripling or doubling year-over-year in the years from 2012-2013. Sales growth slowed down in the U.S. in 2014 but continued apace globally.

Again, this category is a clear win for EVs, particularly when we look down the road a few years at the many additional EVs and PHEVs planned for the market.

Vehicle expense

We don't know the true production cost of EVs or FCVs, unfortunately. We can, however, conclude that costs are coming down rapidly for EVs because some manufacturers are already offering significant cost reductions. GM, for example, in early 2014 took $5,000 off the sticker price of its Volt PHEV, dropping the base model's price from $40,000 to $35,000. The cost savings were made possible mostly from reductions in battery costs.

Time will tell regarding the true cost of FCVs. The companies planning FCV leases in 2014 and 2015 haven't provided significant information on true costs.

In terms of costs to customers (as opposed to production costs), we have more information. Again, the cost of the new Tucson is $499 per month for a three-year lease, plus $2,999 due at signing. The Nissan Leaf is available for a similar lease (without free fuel, which is included in the Tucson lease) for about $199 per month and $1,999 due at signing. The GM Volt has some equally good lease deals available. The Tesla Model S is more pricey, coming in at over $1,200 for a base model lease. There are many other BEVs and PHEVs available at various price points, so it's hard to make any blanket conclusion regarding the customer cost of FCVs versus EVs.

One conclusion we can make: costs are coming down rapidly for EVs as battery costs fall. Battery costs are already falling rapidly with increased deployment of EVs and other uses for battery technologies and are in 2015 already at about $400/kilowatt-hour (some manufacturers reportedly have even lower costs already), down from about $1,200 in 2009. These declining battery costs are translating into declining customer costs for EVs, as both GM and

Nissan substantially cut the cost of their vehicles in 2013 and we're seeing a new crop of more affordable EVs in 2014 and 2015.

FCV costs will also fall as the market scales up. However, with the market still facing all the challenges outlined in this analysis, we may not see much scaling before, say, 2020, if at all. Again, the win in this category goes to EVs.

Fuel costs

What about fuel costs? Hydrogen fuel cost is likely to cost about the same as gasoline. In comparison, the cost for EV fuel—electricity—is about one-half to one-fourth the cost of gasoline and thus one-half to one-fourth the cost of hydrogen. This is because electricity is a cheaper commodity, as well as the fact that EVs and PHEVs are so efficient when compared to traditional internal combustion vehicles. Accordingly, there are no fuel costs savings in driving FCVs. There are major cost savings to be had when it comes to the cost of operating EVs, and this has always been one of the major benefits of EVs.

Another downside to FCVs is that hydrogen fuel evaporates from the vehicle tank when left idle for more than a couple of weeks. While this won't impact most regular drivers, it does pose a downside to those who travel frequently by air and leave their cars for extended periods of time. Fuel evaporation will not only be a hassle for those consumers, it will also add costs because of the losses from evaporation. This issue may be resolved with better technology, and it is not yet clear how big an issue it is in the first generation of FCVs.

Fuel (charge) loss afflicts at least some EVs as well, but the losses appear to be fairly minimal and perhaps are only significant in cold climates or seasons.

Energy storage and grid balancing

Both FCVs and EVs can be used as significant storage and grid balancing tools. The California Public Utilities Commission is currently considering new rules for what is now called vehicle-grid integration or vehicle-to-grid. Even though the CPUC is currently looking at just EVs, much of the same policy and technology structure will apply equally to FCVs.

When their level of market penetration is higher, EVs will present a great opportunity to absorb low-cost electricity and discharge that same electricity back to the grid during higher-price periods. And so will FCVs.

However, because of the large conversion losses in creating hydrogen from electricity and then converting it back to electricity, it would make a lot more sense to use EVs for storage and grid reliability, rather than using FCVs.

The win in this category also goes to EVs.

EVs dominate in most categories

Figure 1 offers a summary comparison of EVs and FCVs, based on one to three checkmarks per category. EVs do twice as well as FCVs. This is, of course, a subjective exercise, but it is based on the fact-driven analysis presented here. It may be the most appropriate way to compare technologies and policies because it recognizes that these issues can be quantified only in an approximate manner. At the same time, it doesn't shy from attempting at least a rough quantification for better policy analysis. Unfortunately, state and federal policymakers don't often engage in such comparisons. Instead, policy decisions are based on experts' or officials' opinions, which generally are offered without such comparisons.

As discussed above, the 2.5 times loss in efficiency for FCVs compared to EVs is surely the biggest issue facing FCVs from a policy perspective. How this issue translate into actual market impacts remains to be seen.

Summary comparison of EVs and FCVs

	EVs	FCVs
Infrastructure cost	x	
Car cost	xx	
Fuel cost	xxx	
Fuel efficiency	xx	
Range		xx
Balancing grid (V2G)	x	
Storage medium		x
Energy density		xx
Fuel availability	xxx	
Re-fueling time		xxx
Fossil fuel reduction	xx	
Disposal/recycling		x
Residual value of vehicles	x	
Momentum	xxx	
Totals	**18**	**9**

In the final analysis, however, we shouldn't rule out FCVs as being a helpful part of the "silver buckshot" solution to wean us from fossil fuels in transportation. While EVs and PHEVs clearly out-perform FCVs in most categories, the fact that FCVs can be refueled in a manner similar to regular cars and that they enjoy a far larger range than most EVs do provide substantial benefits. It is the other issues highlighted in this article that will ultimately tip the balance in terms of consumer acceptance or rejection of FCVs—and only time will tell how that unfolds. My feeling is that the market has already decided that battery electrics are going to be the primary alternative to internal combustion engines, but maybe my crystal ball is cloudy.

Chapter 13

The future of the electric car

There are now over two dozen all-electric and plug-in hybrid electric vehicles on the market today and at least two dozen additional models slated to be available in the next couple of years. Mercedes alone announced in early 2015 that it will be offering ten different plug-in hybrid models in the next couple of years, following a similar announcement by BMW in March of that year. Not to be outdone, BMW shortly thereafter announced that every vehicle it makes will be offered in a plug-in version.

The world's best-selling EV is still the Nissan Leaf, the modest little passenger vehicle, with over 165,000 units sold by March of this year since its release in late 2010. A new study found that the batteries have been very reliable, with 99.99% of the 35,000 Leafs sold in Europe still working perfectly.

The Mitsubishi Outlander PHEV and the Tesla Model S are the second and third best-selling EVs today, even though the Outlander hasn't been available in the U.S. yet.

Global sales of EVs/PHEVs were 320,000 in 2014 alone, an 80 percent rate of growth and on pace to easily exceed 500,000 in 2015. Cumulative sales reached 740,000 vehicles (Figure 1) by the end of 2014 and will top one million around the middle of 2015. This is still less than 1% of the global auto market, but it shows that EVs and PHEVs are here to stay. Global vehicle sales are projected to be about 89 million in 2015. If global EV sales reach 500,000 this will be about 0.6%, putting us on track to reach 1% of global sales in 2016.

As I described in a previous chapter 1% is halfway to market dominance in terms of the doublings required to get from nothing to 1% and then from 1% to 100%. It takes seven doublings to get from 1% to over 100% and the same to go from nothing to 1%.

Figure 1. *Global EV sales through 2014 (source: Centre for Solar Energy and Hydrogen Research).*

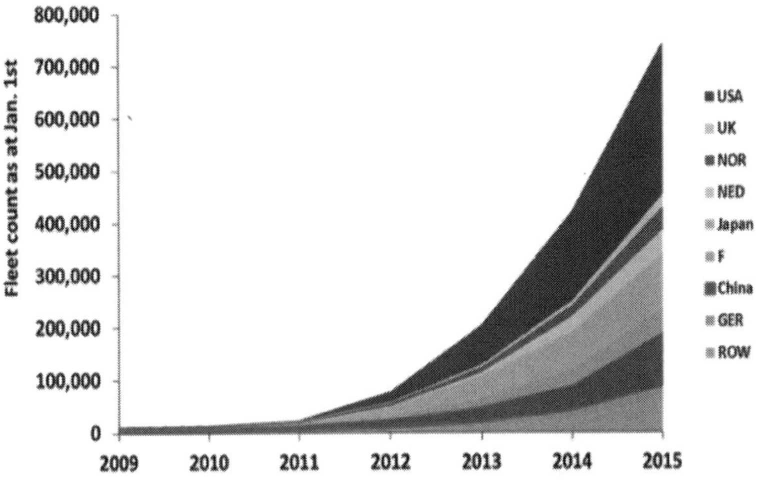

In this chapter, however, I'm not going to talk about the magnitude of sales in the future. Rather, the future I'm focused on here is the future of EV technology itself. How will EVs evolve in the next decade, the years that are already a little bit visible on the horizon? I'll focus on three main trends: 1) improving battery technology; 2) lightweighting of vehicles; 3) automation. Sorry, no (mass market) flying cars are around the corner. Yet.

Improving battery technology

Batteries are an exponential technology so we can gain a lot by looking at learning curve models for batteries. Today's prevailing battery technology for EVs is lithium-ion. It's good but far from perfect. Even though today's lithium-ion batteries hold two and a half times the energy they did in 1991 and cost ten times less, lithium-ion technology is still too energy diffuse and expensive compared to where it needs to be to become truly mainstream. But we're well on the way. Figure 2 shows the various battery technologies in terms of their energy density.

Figure 2. *Energy density of battery technologies (source).*

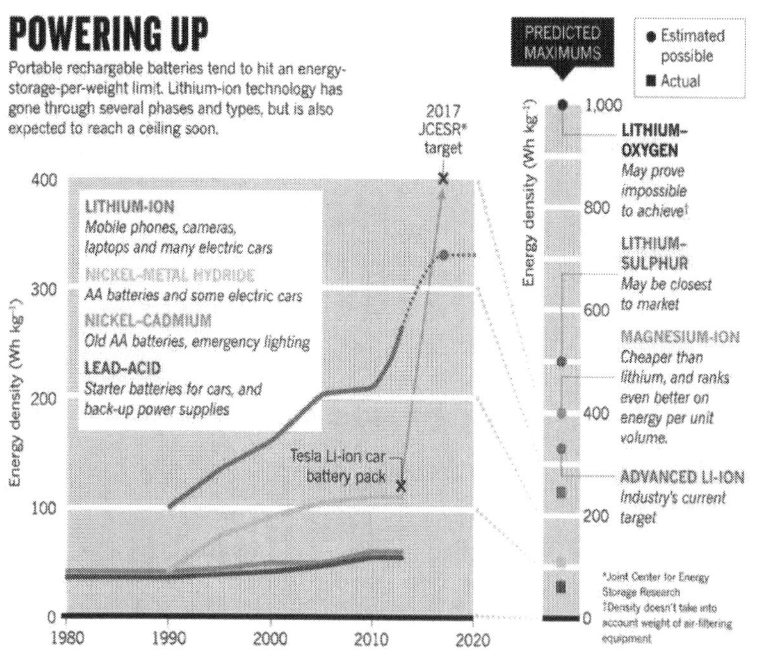

The Holy Grail (for now) for EV batteries is getting costs down to \$100/kWh, down from \$250-300/kWh today. The expectation is that at this cost EVs can compete with internal combustion vehicles without subsidies. The exciting development in the last year is that not only is Tesla's Musk saying they hope to reach this level in the next decade with production from their Gigafactory, but others are agreeing with him, including the Motley Fool, the independent financial research entity.

Lithium-ion batteries will probably start showing marginal improvements around that time, however, in terms of continuing cost declines and increasing energy density. Most experts today think that lithium-ion batteries will top out at around 400 watt-hours/gram or even less, up from around 250 today. Around that point it's likely that we'll see markets shift toward different technologies in order to continue the trend toward increased energy density.

One new technology that shows promise is lithium-air (also known as lithium-oxygen), which uses ambient air to aid energy flows and results in less degradation than in other types of batteries. The long-term Holy Grail of lithium-air batteries is achieving an energy density on par with gasoline, which is theoretically achievable with this technology, and would allow batteries to take up much less space than today. This is about ten times the energy density of today's batteries.

The future is not all clear, however, for lithium-air batteries and some researchers think they're a lost cause. Others aren't ready to bow out yet, including Peter Bruce with the University of St. Andrews in the United Kingdom: "We are closer to what's needed than we were a few years ago."

As with Moore's Law and the speed and cost of computing power, we can expect to continue to see major improvements in battery technology in the coming decade. With Tesla and Panasonic throwing literally $billions behind improved lithium-ion batteries it's likely that lithium-ion will remain the go-to battery for EVs in the next decade. It's also possible, however, that other types will make major breakthroughs in that same time period and reach the broader market.

My best bet is that we'll continue to see incremental improvements in lithium-ion energy density but continue to see steady decreases in costs. This is the case because costs will decline based on increases in the scale of production and learning curve effects relating to this scale and associated manufacturing techniques, independent of any major improvements in the actual architecture of the batteries. Such improvements will be icing on the cake but probably not necessary to achieve major improvements over today's density and costs.

Lightweighting

Interestingly, we can achieve exactly the same improvement in EV range or reduction in costs by making cars lighter as we can by making more energy-dense batteries. And it may actually be easier to do it by making cars lighter because there is less of a diminishing return phenomenon in taking today's relatively heavy cars and making them lighter with the use of new materials and techniques.

We're already seeing many technologies leading to lighter and more efficient cars more generally, regardless of whether the cars are electric or gasoline or diesel. Some kind of fuel economy standard is in place now for 80 percent of the world's passenger vehicle market, according to the International Council on Clean Transportation. Figure 3 shows the vehicle fuel efficiency standards for the world's largest economies at the end of 2014. The general trend is a fifty percent reduction in the emissions from passenger vehicles between 2000 and 2025. We're about halfway to achieving that goal and it is realistic to expect that the 2020-2025 goals will be met due to the many new technologies arriving that will improve fuel efficiency and decrease GHG emissions.

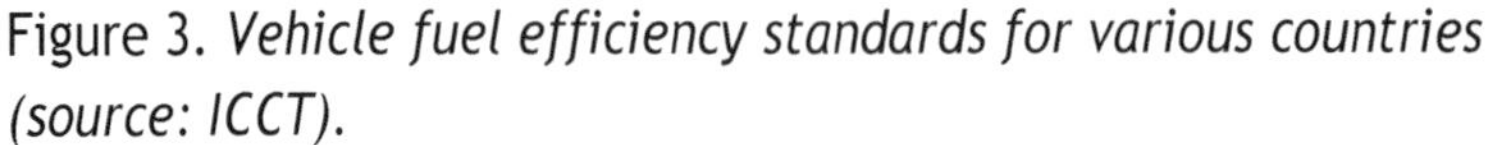

Figure 3. *Vehicle fuel efficiency standards for various countries (source: ICCT).*

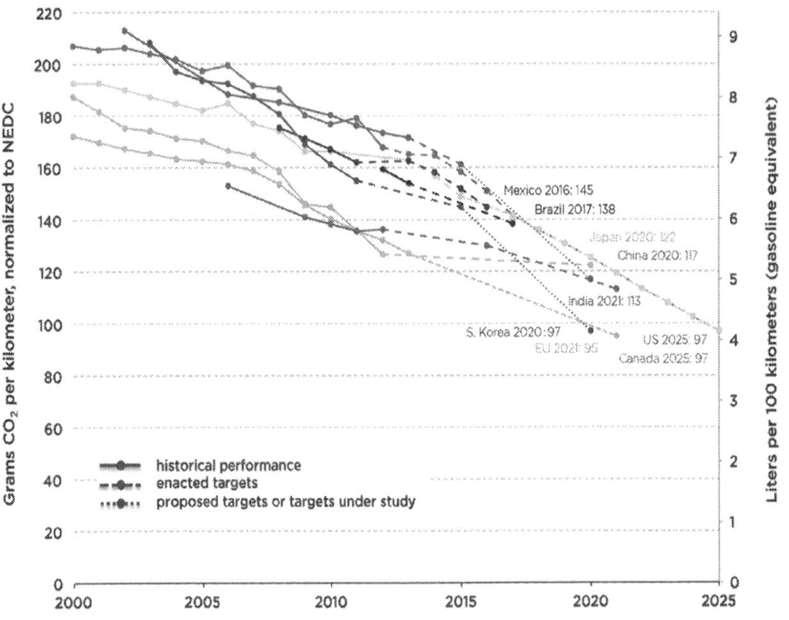

As a consequence of these trends (and many others), IEA projects in its 2015 World Energy Outlook that oil demand will rise to 104 million barrels per day (mbpd) by 2040, up from 91 mbpd in 2014. This is a significant increase but it's a large decline from previous forecasts that reflects the increasing role of improved vehicle fuel efficiency around the world. We can anticipate additional vehicle fuel efficiency to materialize in the coming decades, going far beyond the 54.5 miles per gallon requirement under the U.S. CAFÉ standard for 2025.

Many EVs are already quite small. The smallest commercially available model is the Renault Twizy, with over 15,000 already on the road in Europe (it's not available in the U.S.). The Fiat 500e is tiny (I should know since I drive one), and the Smart ForTwo EV is even smaller. Are we destined for a future of electric go-karts? Well, we'll surely have many more of these truly tiny vehicles on the road, particularly in dense urban areas, but we're also going to see plenty of normal size cars becoming increasingly lightweight.

Amory Lovins and the Rocky Mountain Institute have for many years pioneered the "Hypercar" idea that includes a radical redesign when it comes to weight. Lovins describes many aspects of this vehicular evolutionary process in his 2011 book, *Reinventing Fire*. He states: "Investing R&D effort in vehicle fitness (which got about 100-fold less in U.S. research budgets through 2010) will yield the same result [as better batteries] with less cost, time, and risk."

Ford made big news recently by announcing its use of aluminum for much of the frame in its top-selling F-150 line of trucks. The 2015 model is 700 pounds and about 15% lighter due to this shift from using steel in favor of using aluminum. Ford is claiming up to a 20% improvement in fuel efficiency due largely to this change in weight. The aluminum used in these trucks is about twice as strong as steel and thus safer in accidents as well as lighter.

Steel is not giving up without a fight, however, and some companies are planning to use high-strength steel to lightweight cars rather than alternatives like aluminum. ArcelorMittal, a steel manufacturer, released a survey suggesting that pickup trucks could achieve an average 23% improvement in weight by using the new high-strength steel products combined with expected improvements in power trains.

One idea that Lovins has highlighted for years is also starting to see reality: carbon fiber vehicle bodies. The BMW i3 compact hatchback is far lighter than comparable cars because, in part, its body is made mostly of carbon composites, the same stuff that modern airplanes are increasingly being made from. This is the first car (let alone EV) to use primarily carbon fiber for its body. Carbon fiber is 30% lighter than aluminum and actually stronger. The i3 weighs 2,635 pounds, significantly less than the average weight of 3,000-4,500 pounds for the typical compact car. The Chevy Volt, for example, another highly fuel-efficient car and also officially a compact car, weighs 3,786 pounds. Even the tiny Fiat 500e weighs more than the i3, at 2,980 pounds.

The i3 represents a big shift in long-term trends in car weight because after a massive drop in the 1970s, in response to the oil crises, most vehicles steadily gained weight from the 1980s through the to the present. We can expect, with the federal CAFÉ standards for 2025 being quite aggressive, that other automakers will follow BMW's lead in terms of using new materials in a long-term quest to create much lighter cars.

Over 20,000 i3s have been sold worldwide since its late 2013 release and by all accounts it's a great car with no issues relating to its carbon frame. U.S. sales of over 1,000 vehicles in February of 2015 put the i3 at third place after the Nissan Leaf and Tesla Model S and well above the Chevy Volt and Prius Plug-in hybrid.

Automation

We're also going to see cars in general become far more automated. It seems that the hurdles to driverless cars are now largely legal rather than technical and my feeling is that we'll have fully autonomous cars in the next few years and maybe even sooner for some models.

The environmental benefits of automation (let alone benefits such as gained time) are potentially enormous. A 2015 study from the Lawrence Berkeley National Laboratory found that driverless cars could reduce greenhouse gas emissions from cars by up to 90 percent, with much of those reductions coming from "right-sizing" of cars based on the needs of each person calling up an automatic vehicle. For example, if you just need a quick ride into town a tiny one-person driverless car would speed to your door and toot its little horn to let you know that it's arrived. And if your whole family is going to the beach, well then a minivan with robot driver will show up at your door.

Tesla has earned a lot of attention when it comes to driverless cars and they may well lead the pack to achieving a fully autonomous vehicle. Musk has announced a suite of "autopilot" capabilities that are coming via software update this summer to the company's signature Model S. Musk stated that the new features could in theory allow the car to drive itself from parking lot to parking lot on long drives throughout much of the U.S., but for now it will be limited to highway driving due to the legal obstacles currently in place.

Many other brands are also working on fully autonomous cars and most luxury cars have an increasingly sophisticated suite of automated driving abilities. The Audi A7 completed a test drive with its autopilot from the Stanford campus in Palo Alto to Las Vegas earlier this year, without incident. Mercedes and Cadillac models also have many autopilot features already included, with surely many more features to come.

In sum, it seems pretty likely that we'll have fully autonomous cars here by 2020 or even sooner. That's pretty cool and I personally look forward to telling my car to take me to Yosemite for the weekend as I kick back for a nap.

Chapter 14

The almost perfect solution: driving on sunshine

In thinking about possible pathways for a low or zero carbon economy, Shangri La could well be a solar-powered transportation system, with electric cars literally running on sunshine. That future may, however, already be here.

A recent survey found that 32% of electric vehicle (EV) owners in the western U.S. have solar panels (PV) on their homes. We don't know how much of the miles driven come from those panels because they may not be sized appropriately to supply the home demand and the vehicle demand. Regardless, it is clear that more and more homeowners will, as prices for PV and EVs continue to fall, opt for what we can call the PV4EV solution.

Even though a PV system can still cost a pretty penny today the savings from driving on sunshine versus driving on fossils can make up for the initial cost in a decade or sometimes far less. In crunching the numbers for various models I found that payback times for the PV4EV solution are generally around ten years—good but not great, at least from a purely economic perspective. As costs for solar continue to fall and gasoline costs continue to rise, this payback will become shorter and shorter.

To drive 30 miles a day (about the average distance driven each day by Americans) requires from 1.9 to 3.2 kilowatts of PV for the four models included below. Figure 1 shows the PV system size needed just to run the vehicle, not for any additional home power use.

Figure 1. *Info on four EVs available today.*

	Miles per kWh	Battery size (kWh)	Cost per mile	PV (kW)* needed for 30 miles per day
Smart ForTwo Ev	3.8	16.5	$ 0.04	1.9
Leaf EV	3.1	24	$ 0.05	2.4
Volt EV	2.3	16.5	$ 0.07	3.2
Tesla EV	3.5	60	$ 0.04	2.1
* Assuming 4.5 hours average production per day and 90% system efficiency				

Using this data from Figure 1 we get 7-12 years payback times for the PV4EV solution for the four models examined (Figure 2). Perhaps surprisingly, the Volt has the shortest payback time, at 7.7 years. Tesla's Model S, also surprisingly, has the second shortest solar payback. This is because it is a highly efficient vehicle and I'm assuming that Tesla drivers are going to drive more each year than a Smart or Leaf owner (15,000 miles versus 10,000 miles)—just because it's so much fun to drive a Tesla and the fueling is free.

Using EPA figures for miles per kWh we get the payback times in Figure 2, ranging from 8-12 years, with the Volt faring better in comparison to the EVs, largely because the Volt is programmed to be more cautious in using its battery capacity than the EVs (less of the battery is available to the driver even though the official capacity is 16.5 kWh.

Figure 2. *Payback times using EPA figures for m/kWh.*

Item	EV cost*	Solar panel cost	5-year car fuel cost savings**	Solar payback time (years)
Smart ForTwo Ev	$ 15,750	$ 8,102	$ 3,500	11.6
Leaf EV	$ 18,980	$ 7,625	$ 3,500	10.9
Volt EV	$ 24,185	$ 8,102	$ 5,250	7.7
Tesla Model S	$ 59,900	$ 9,259	$ 5,250	8.8

After the PV system is paid for in fuel savings the homeowner enjoys literally free power for the remaining life of the system, which is warranteed under California rules to produce at least 80% of full capacity after 25 years. That's a good deal in my book.

I haven't considered any cost premium for the EV purchase in my PV4EV calculations because each car is essentially a high-performance vehicle, when compared to its fossil-powered equivalents, so the cost is justified by the value. This is particularly the case when we include, as I have done, available rebates and tax credits. There is a societal cost, of course, to these subsidies, but the cost is minuscule compared to the potential benefits of accelerating the transition away from fossil fuels.

These subsidies won't stick around forever and nor should they (both have phaseouts already built in). However, it seems very likely, given recent price reduction trends, that these EVs will continue to represent a good value proposition even as rebates and tax credits fade away - particularly when we consider the cost savings from driving on sunshine versus fossils.

My cost estimates for PV don't include any state rebate, because that is going away quickly for most Californians. Fully 58% of solar systems were installed in the first three months of 2014 without any state rebate. My cost estimates for solar do, however, include the federal 30% tax credit available for solar through 2016.

Solar PV costs continue to plummet, so my calculations will continue to become more favorable over time. See Figure 3 for the recent history of solar cost reductions through the 1st quarter of 2014. The most recent report from the SEIA market insight report shows that utility-scale solar (and other types too) continued their amazing cost declines, averaging just $1.55/watt in the fourth quarter of 2014, down from $1.85/watt in the first quarter. These trends show that an all-in installed cost of just $1/watt is probably just a couple of years away.

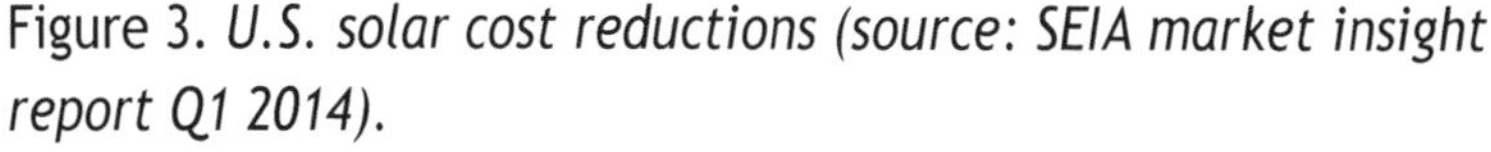

Figure 3. *U.S. solar cost reductions (source: SEIA market insight report Q1 2014).*

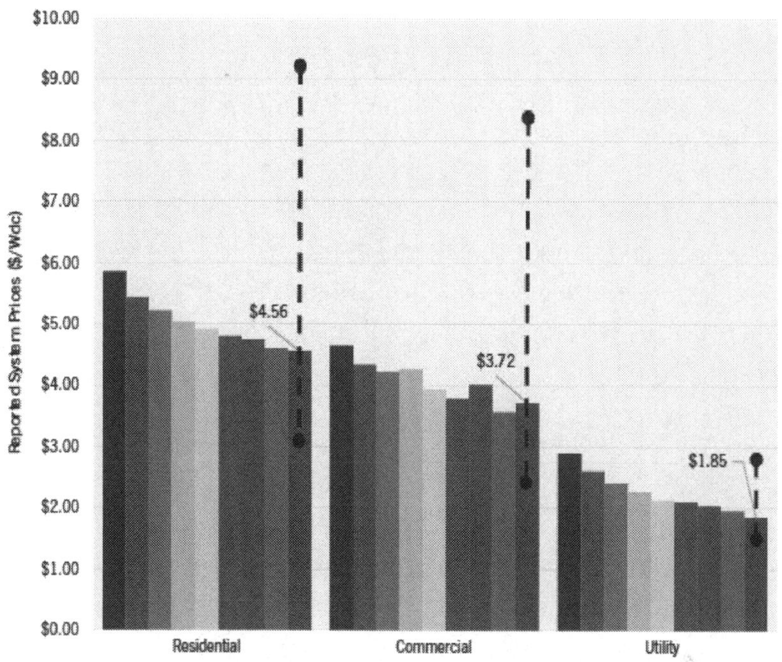

The Santa Barbara Community Environmental Council (my former employer) has a number of engaging case studies of people choosing to drive on sunshine. Based on the numbers above, as well as the obvious environmental benefits of PV4EV, we can expect many more people to start driving on sunshine instead of fossils.

Chapter 15

What about energy storage?

Is energy storage the new holy grail of a clean energy future? Energy storage can take the form of large batteries, flywheels, pumped hydro facilities, compressed-air energy storage, and many others. However, the exciting "new kid on the block" is battery storage, largely because of its relatively small size, modularity and rapidly declining costs.

California is at the forefront of the utility-scale storage revolution and policymakers have recognized the potential role of energy storage in a number of new programs. The biggest of these programs is the AB 2514 energy storage procurement program. This law required the California Public Utilities Commission (CPUC) to consider creating an energy storage procurement mandate for utilities. And that's about all the law did, leaving most issues to the discretion of the CPUC. The law did, however, require that any storage that is mandated must also be cost-effective. That is, the mandate couldn't lead to any net costs for ratepayers. (This is a common requirement in California for new programs, despite the common myth that legislators and policymakers don't care about costs.)

The CPUC took up the challenge that Sacramento offered with AB 2514, and in late 2013 (in Decision 13-10-040) the CPUC created a new program requiring utilities to procure 1.325 gigawatts of new storage by 2024. This is a lot of storage, enough to produce power at peak output for over 1 million California homes, as well as to provide numerous other grid benefits at the same time.

Under this new program, the utilities must issue requests for offers every two years starting in December 2014, with defined megawatt targets for each RFO. The targets are divided into three "buckets": 1) transmission-interconnected; 2) distribution-interconnected and 3) behind-the-meter projects such as electric vehicles or home battery systems for solar.

The CPUC deferred ruling on the issue of cost-effectiveness until the utilities submit their energy procurement program rules for consideration by the Commission. The utilities submitted these applications in late February and the CPUC approved them later in 2014.

Time will tell how the CPUC rules on cost-effectiveness since at the time of writing (early 2015) the CPUC had still not ruled on this issue. Even though many parties do have concerns about the details of this major new program, it is fairly certain that over the next ten years, this program will be a strong boost to the nascent energy storage market in California and, by extension, in many other jurisdictions in the U.S. and abroad.

The hope, of course, is that by providing a strong market signal to producers and developers, the price for storage technologies will come down further. As discussed in previous chapters, we've seen strong price declines in solar power and many other renewable energy technologies in the last few years. Battery technologies have also come down significantly in cost in the same timeframe. There is still, however, much room for further declines as production scales up. Figure 1 shows that most storage technologies are either in the early deployment stage or still in R&D. Early deployment technologies like lithium-based batteries are promising, and the hope is that new programs like those in California will promote rapid market maturation and bring costs down dramatically, as we've seen happen with solar and wind power.

Figure 1. *Stationary Energy Storage Technology Maturity (source: IEA Energy Storage Technology Roadmap 2014).*

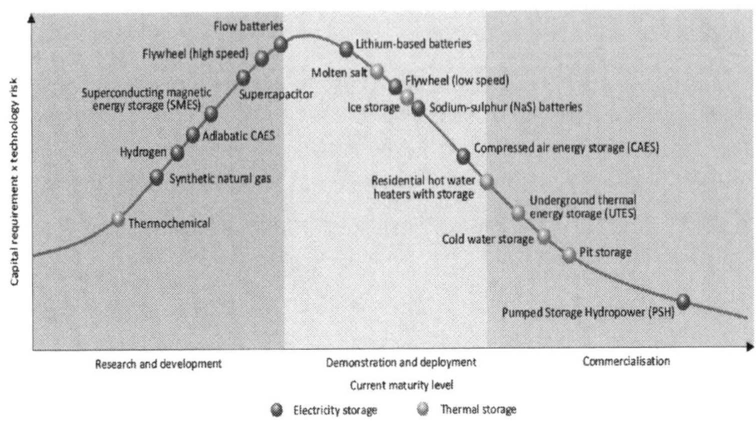

California has two other programs relevant to the energy storage market: SCE's Local Capacity Requirements RFO and the Self-Generation Incentive Program. It is likely that any storage procured in these programs will count toward the 1.325-gigawatt goal of the energy storage procurement program, but these other programs will entail different types of RFOs.

SCE's Local Capacity Requirements RFO

Southern California Edison's LCR RFO is designed to ensure that sufficient local power is available in the West L.A. Basin. The program extends through 2020, and SCE will offer a number of RFOs during this period. This program is open to all types of energy production, including renewables (except nuclear and coal, which aren't allowed in California), but also demand response, energy efficiency and energy storage technologies. In fact, the CPUC required SCE to procure 50 megawatts of energy storage via this RFO by 2020.

My company submitted a bid to SCE under this RFO in December 2013, for a 2-megawatt (8 megawatt-hour) battery project, on behalf of the L.A.-based company Absolutely Solar. The project was to be owned by Absolutely Solar, but SCE would operate the facility as a grid reliability asset. It would connect to the distribution grid and provide local energy and resource adequacy, as well as other grid benefits. Our project didn't get short-listed by SCE, but we learned a lot about the process and we look forward to submitting additional bids in the future.

Self-Generation Incentive Program

The Self-Generation Incentive Program (SGIP) is solely for behind-the-meter storage projects and provides a generous rebate to owners of new energy storage systems. Only a few megawatts of storage are on-line under this program for now, but interest in the rebate is growing. The utilities are also now more incentivized to work with power customers to install batteries under the SGIP because these projects count toward the utility's behind-the-meter storage mandate under the AB 2514 energy storage procurement framework.

So, are we facing a tsunami of energy storage in California and the U.S. more broadly over the next decade? My feeling is that we'll see plenty of opportunities to bid for contracts to sell storage capacity to the utilities, under the programs mentioned here, as well as possible new ones. The timeframe for development of projects is far less certain, due largely to the long timeframe that the utilities have to develop these projects (up until 2024 in most cases), along with various flexibility mechanisms provided by the CPUC. These mechanisms allow the utilities to defer procurement until later years in the program under certain circumstances.

Again, time will tell how the details unfold. But we are in for an interesting time as these programs develop. While the timeframe for development of these energy storage programs is uncertain, it is clear that we will need a significant amount of storage to firm up renewables like wind and solar as these technologies come to dominate the grid in coming decades.

Chapter 16

What happened to peak oil?

Peak oil is the point at which global oil production peaks and can only go down. When this happens it will lead to a slew of negative economic impacts because our economy is built in large part on energy, primarily fossil-fuel energy. I've followed the peak oil debate for a decade and written numerous articles on this issue. The good news now is that there's a good chance we will see peak oil *demand* arrive before a permanent peak in global peak oil production induced by physical limits.

This chapter looks at the rise and fall of the peak oil debate itself and offers some thoughts on where we are in terms of peaks in global production and demand. From 2005 through about 2010, the energy world was atwitter with discussions of peak oil. But in the last few years, as U.S. oil production has dramatically ramped up, many peak oil believers have been left looking a bit silly.

I became really worried in the 2007-2008 timeframe about possible major problems from the earlier-than-expected arrival of peak oil because of the massive run-up in oil prices at that time. It was only the economic crash that began in mid-2008, the largest since the Great Depression, that led to a decline in oil prices from their record peak of $147 a barrel in July of 2008. Oil prices plummeted at that time to $33 a barrel—more than a 75% reduction—in early 2009 before climbing steadily back to over $100 a barrel in 2011 and staying around that level until late 2014.

This sustained high price was the highest-priced five years for oil in U.S. history and it seemed to indicate a long-term structural shift in oil supply and demand, and thus higher prices. If there was plenty of oil to meet demand, as the "cornucopians" argued, why were prices staying so high? As the global economy climbed out of the Great Recession that began in 2008 it faced the strong headwinds of high oil prices. Because oil is such a large source of energy, price swings have big impacts on the economy. For example, every $10 shift in oil prices is equivalent to a 0.1-0.3% impact to the U.S. economy. So the increase in prices from 2007 to 2008, from $56 to $147, was equivalent to a 1-3% GDP hit. This may have been enough to induce the Great Recession all by itself but even in light of the far better-known financial crisis of 2008-2009, we can fairly say that oil prices were a big factor that got overshadowed by the financial crisis.

Why were prices so high until late 2014? Were we on the "undulating plateau" (to use a phrase made famous by Daniel Yergin and his company IHS) of peak oil or was something else going on? Global supply was indeed constrained for a number of years and spare capacity was limited. This allowed disruptions in Libya, for example, in early 2011 to lead to a big price spike. Before there was such tightness we wouldn't have seen much of a price swing from Libya's disruptions.

This is a complex debate but I don't mind admitting that I was very wrong on at least one aspect of this debate: I didn't expect to see, here in 2015, the U.S. vying with Saudi Arabia and Russia to be the world's biggest oil producer. The unconventional oil revolution here in the U.S. took a lot of people by surprise, including me. The peak oil crowd has been silent for a while now because all those people were also surprised by the unconventional oil production spike in the U.S. and a slower-than-expected decline in conventional oil production. A lot of people who have followed the peak oil debate peripherally assume that the peak oilers were simply wrong, way wrong, about peak oil.

One thing at least is clear at this point in time: the U.S. has achieved a massive increase in oil production (crude oil and condensate) since about 2009, increasing from about 5 million barrels per day (mbpd) to over 9 mbpd by early 2015. This increase has brought our country almost on par with Saudi Arabia and Russia as the world's biggest producers. When we add in other liquids production like natural gas liquids, refined products, and ethanol the U.S. is over 14 mbpd and leads the world as a liquid fuel producer.

Figure 1. *Comparing U.S., Saudi, and Russian crude oil and condensate production (source: EIA).*

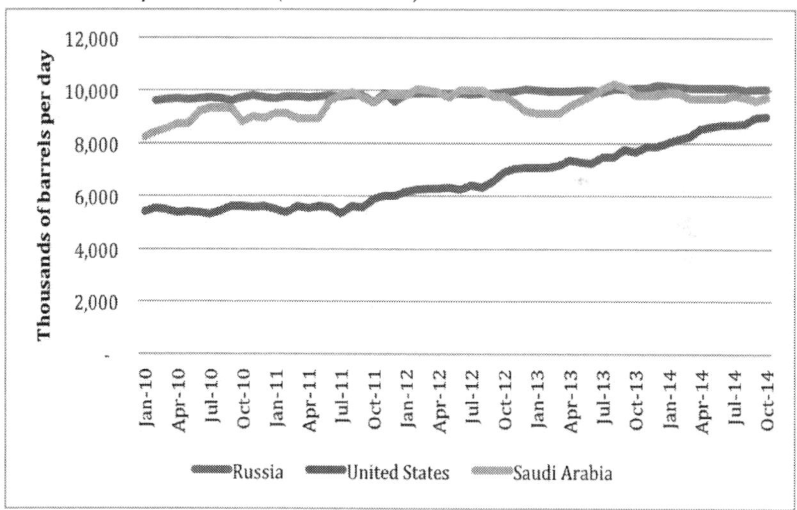

Why is this such a big deal? Well, it shows that the U.S. was able to turn around a multi-generational decline in oil production with new technologies and it does give pause to even the most cautious peak oilers because we now have very good proof in the pudding that oil production can change, and change big, in just a few years. Maybe there is nothing inevitable about long-term oil production declines anymore. To illustrate this point Figure 2 shows U.S. oil production over the last hundred and fifty years, showing the 1970 oil production peak.

Figure 2. *U.S. oil production since 1860, in barrels per year (source: EIA).*

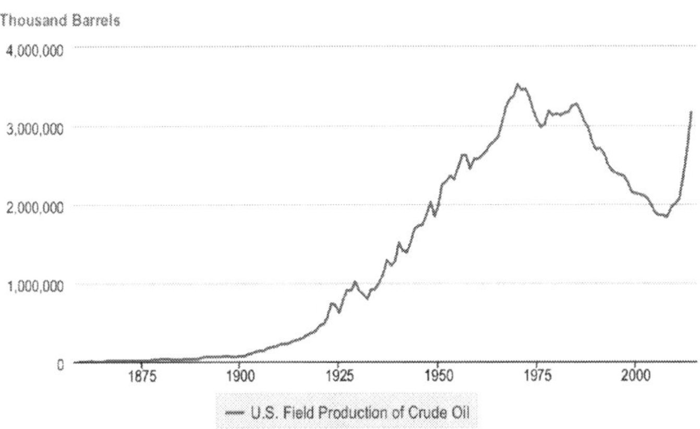

Globally, however, the picture is different. Fracking and horizontal drilling—the technological breakthroughs that led to the U.S. production resurgence—haven't caught on around the world in the same way that they have in the U.S. Global production is up from the plateau years of 2005-2010, but by only about 5 mbpd. This is, in fact, not much larger than the U.S. increase in production (about 4 mbpd) so the *global* increase is largely equivalent to the U.S. increase in production and demonstrates that global production other than the U.S. has been on a plateau for the last ten years, at about 85 mbpd.

Figure 3. *Global oil production 1980-2013 (source: EIA).*

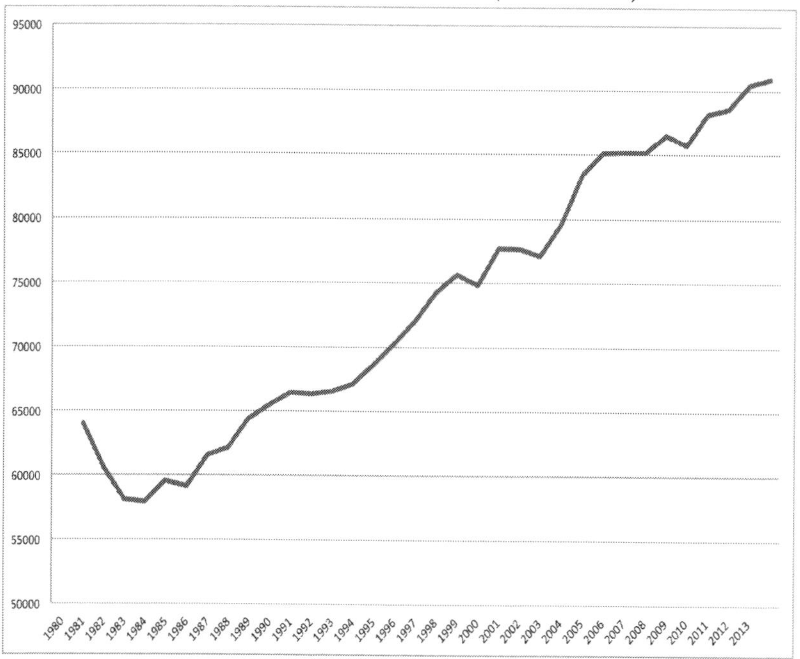

In light of these new facts, let's examine some of the arguments behind peak oil and see if there's still any relevance to this debate. Or do all of us peak oilers simply need to eat a healthy portion of crow and believe more in the free market to incentivize new drilling technologies and new areas for production?

The world wakes up to peak oil?

The 2008 World Energy Outlook issued by the International Energy Agency was the document that really got me on the peak oil bandwagon. This is an annual report and IEA finally started listening to the ever-growing peak-oil crowd and completed a supply-side analysis in 2008 of the world's 800 biggest oil fields. In previous analyses, IEA had projected oil and other fossil fuel demand based on economic modeling and simply assumed (literally) that supplies would meet this projected demand. The 2008 WEO used a very different approach. Rather than simply assuming supplies would meet demand, IEA looked at the largest oil fields and calculated their rate of production and decline (this is what a "supply-side analysis" means). They found that these fields were declining far faster than previously assumed, about 7% per year rather than 3.5% per year.

Based on this decline rate, IEA calculated that the world would need eight new Saudi Arabias by 2030 to meet projected demand. In the medium-term, it projected a likely "oil supply crunch" by 2015 because we'd need three to four new Saudi Arabias by then to meet projected demand. We know now, of course, that no supply crunch came by 2015. To the contrary, we have seen at least here in the U.S. a major swing back to increased production (Figure 2).

At the time of this writing the West Texas Intermediate price was under $50 a barrel and Brent was priced around $60 a barrel. I won't be at all surprised, however, if these prices double before the end of 2015. This is the case because we're already seeing the number of oil rigs in the U.S. plummet in response to these much lower prices.

Figure 4. *Active oil rigs vs. production (source Bloomberg.com).*

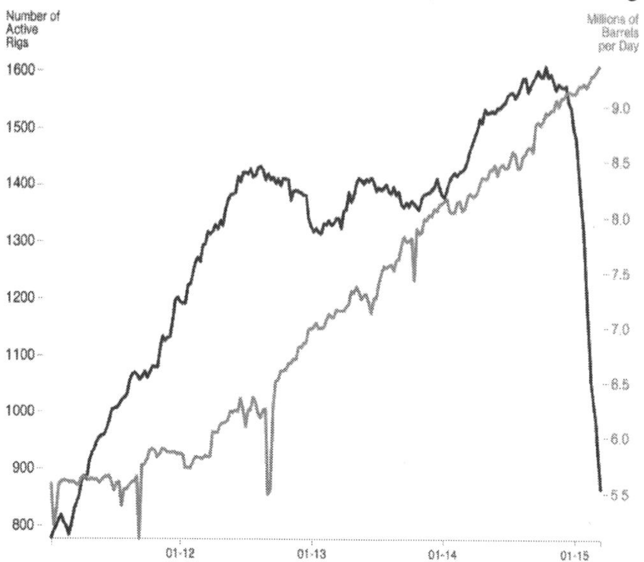

The coming dip in production will not, however, be driven by physical limits on production capacity. Rather, the drop will be due to the lack of economic viability of unconventional oil in the relatively low price environment we're seeing in early 2015. Even though this is a real limit, the difference between this kind of economic limit and physical resource depletion is that once prices rise again we'll surely see a lot of the retired production come back online relatively rapidly. What it does illustrate, however, is the very real trend toward higher production costs as the "easy oil" is depleted around the world.

We'll see below, however, how the new global oil dynamic may be one of peak *demand* rather than physical limits on production. But any peak in global demand will depend very much on the degree to which EVs and other petroleum use-reducing technologies are adopted. It's too early at this point in the game to say how this will unfold.

What do we know with certainty with respect to peak oil?

While the timing of peak oil (whether in terms of physical production limits or peak demand) will always be uncertain, there are a few things about this debate that we can know with relative certainty. For example, as just mentioned we know that oil is getting harder and harder to find, which is why new drilling techniques are being invented and so many new drilling platforms are going into deeper and deeper water and more and more remote areas of the world.

We know that the oil majors are spending more and more to produce less oil and add less and less to their oil reserves. A February 2015 Reuters analysis found the following: "Over the past decade, the biggest Western oil companies have seen reserves growth stall, production drop 15 percent and profits fall by almost a fifth—even as oil prices almost doubled, a Reuters analysis of corporate filings shows." Similarly, the respected industry news site, oilprice.com, ran an article in early 2015 that described a case for a possible major increase in global oil prices in the next decade because 2014 was the record fifth year in a row that oil majors failed to replace their produced oil with new discoveries. "Despite record levels of spending, the largest oil companies are struggling to replace their depleted reserves."

We also know that unconventional oil wells have far higher production declines than conventional wells. Oilprice.com also reported on a new analysis by David Hughes and the Post-Carbon Institute that predicts an earlier and stronger decline for U.S. oil production resulting from far higher than normal decline rates in new wells. The average decline rate for conventional wells is about 5% per year. In stark contrast, the average decline rate for fracked and other unconventional wells is 60-91% over the first three years, at which point declines slow down in an extended but low level of production. See Figure __ for an example of average decline rates in the Bakken formation in North Dakota. The obvious result of these high decline rates is the need to drill more and more new wells to make up for declining production.

Figure 5. *Post-Carbon Institute analysis of average decline rates for tight oil production in the Bakken formation.*

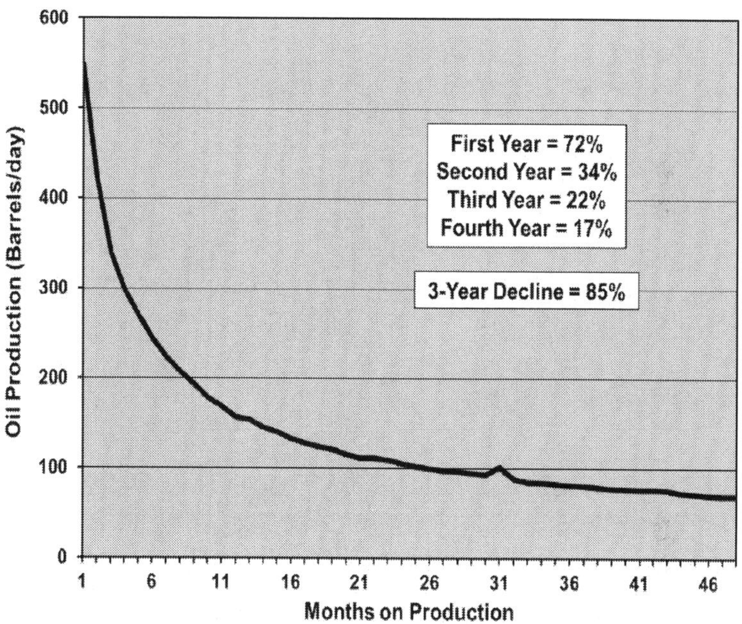

We also know that oil-producing nations often see steadily rising demand for oil combined with long-term production declines, leading eventually to zero oil exports from these countries. This is a phenomenon that has happened already with dozens of former oil-exporting countries. I'm going to focus on this issue for the rest of this chapter because it seems to me to be probably the biggest issue with respect to the peak oil discussion. Even if global oil production continues upward for some time, for oil-importing countries like the U.S. and almost all of Europe, what matters is not global production but global net exports—available oil, in other words. Even with the resurgence of U.S. oil production we still import about five mbpd (about 25% of our demand), equivalent to about half of Saudi Arabia's entire exports, though U.S. imports have fallen dramatically, from as much as 13 mpbd in recent years (Figure 6). I focus on Saudi Arabia next in order to highlight the net exports issues. As U.S. demand increases and our oil production plateaus and declines it's likely that we'll see our net imports figure rise substantially again.

Figure 6. *U.S. net oil imports and petroleum products (source: EIA).*

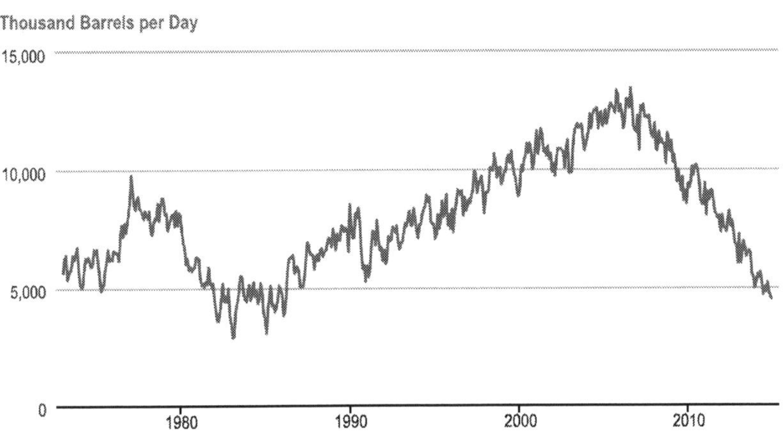

The Saudi Oil Problem

Saudi Arabia's national oil company, Saudi Aramco, pumped about 11.5 million barrels per day in 2014, up from about 9.5 million in early 2009. The Saudis are pumping more oil now than they have in decades, along with the rest of OPEC, which is at almost a 25-year high for combined oil production.

Russia held the top spot for oil production for a couple of years, but Saudi Arabia has come roaring back since 2010. The U.S. is in third place with about 9 mbpd.

Net oil exports, are, however, a different picture. The U.S. was for some time the world's biggest importer of oil, before being surpassed in 2014 by China. While our production of oil has taken an unusual upward tick in the last few years, spurred by record high prices and new drilling technologies, and our consumption of oil has declined even further due to increased energy efficiency, conservation and a still-struggling economy, we still import about 25% of the oil we consume.

Saudi Arabia exports about 8 mbpd, with Russia not too far behind at about 7 mbpd.

So far, this is all fairly familiar data. However, what is not well known is the degree to which Saudi Arabia's massive oil exports are threatened by its demographics and a probable decline in its aging supergiant oil fields, particularly from 2020 onward.

A 2011 report from the U.K.'s Chatham House examined this problem in detail. They concluded that Saudi Arabia's oil exports will peak around 2020 and, under current policies, decline to zero by 2038. You read that right: decline to zero. This decline will occur due to the dramatic growth in consumption by Saudi Arabia's rapidly growing population and increases in per capita energy consumption. Saudi domestic consumption of oil is growing at about 7% per year, which leads to a doubling of consumption in just ten years.

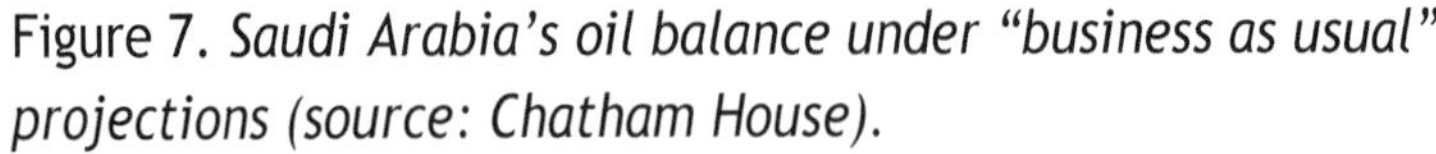

Figure 7. *Saudi Arabia's oil balance under "business as usual" projections (source: Chatham House).*

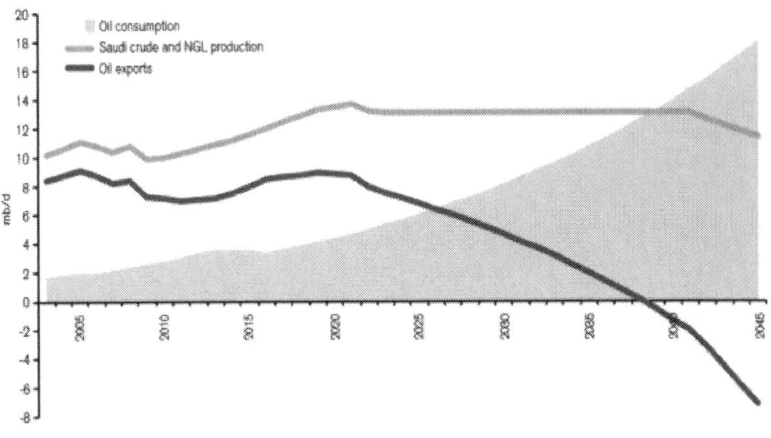

Saudi Arabia's growth in consumption in the last few years hasn't been as high as the Chatham House report forecast. EIA has data through 2013 and the growth in consumption for 2010-2013 averaged 5.0% rather than the 7% the report forecast. However, the 7% figure is an average growth rate through the 2045 forecast period. As is clear from the previous chart the oil consumption curve heads up steadily from 2016 onwards but Chatham House actually forecast a slight decline in consumption from 2012-2016.

Saudi Arabia's problem is not, of course, unique to Saudi Arabia. It is a global problem that afflicts most oil-producing countries because wealth from oil spurs economic growth spurs increased consumption of oil. Jeffrey Brown and Samuel Foucher, an oil geologist and mathematician, respectively, have developed an "Export Land Model" to predict how the global net oil export situation will unfold in coming years. Brown's analysis through the end of 2012 found that the top 33 oil-exporting countries will steadily see their exports shrink to a ratio of about 2:1 for production vs. consumption by the early 2030s. And by 2030, Brown, projects under current trends that *all available net exports will be consumed by China and India alone due to their strong rates of economic growth.*

Brown's net oil export projection declines are due to the demographic explosion that the Chatham House report focuses on but also to declines in oil production in exporting nations. Chatham House chose not to discuss the role of declining oil production, choosing instead to believe Saudi projections of steady or growing oil production, but declining oil production is a real and extremely serious corollary to the demographic explosion. It's also the reason why Brown and Foucher project Saudi Arabia and other oil exporters going to zero faster than the Chatham House report.

Can the shale oil revolution take off outside of the U.S.?

In sum, it seems that the debate about peak oil—to the extent that it still exists today—rests on the degree to which fracking and horizontal drilling techniques can be used outside of the U.S. to expand global production, and the degree to which potential increases in production can offset the declining global net oil exports. Iraq is also a big "x factor" in terms of its ability to ramp up production. By all accounts it has the resources to ramp up production substantially, but with the turmoil being wreaked by Daesh in that country since 2014 it is highly unlikely that Iraq will ride to the rescue in time to make up for the coming decline in U.S. unconventional oil production.

The perils of assuming that other shale oil resources are suitable for oil fracking is illustrated well by the early 2014 downgrade of oil resources in California's Monterey Shale formation. EIA projected before this downgrade that the Monterey Shale contained almost 14 billion (with a "b") barrels of recoverable oil, a large share of the nation's remaining oil resources. The 2014 revision reduced this figure by 96%, in line with a new survey by the U.S. Geological

Survey. Similar downgrades have taken place with respect to shale gas from the Marcellus Shale in the Eastern U.S. (80%) and, very on point with respect to the ability of the shale and fracking revolution to grow legs outside of the U.S., Poland (a whopping 99%). The Post-Carbon Institute report summarizes these three downgrades in Figure 8.

Figure 8. *EIA downgrades of shale oil and gas resources (source: Post-Carbon Institute).*

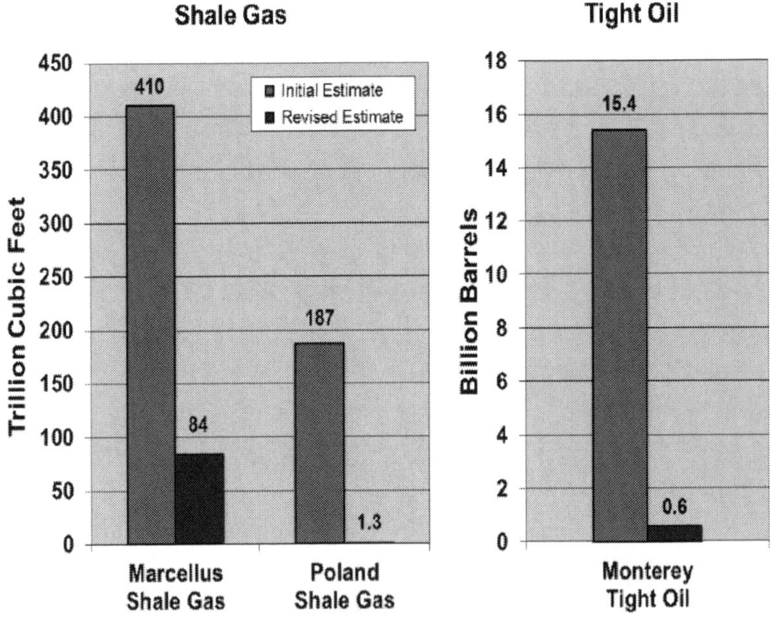

EIA and IEA project, in their 2014 reference (base) cases, that U.S. unconventional oil production will peak around 2020, with total production leveling off at around ten mbpd through the early to mid-2020s. If the fracking revolution can't be exported outside of the U.S. it is likely that we'll see global production decline in the same timeframe: the early 2020s. It is this dynamic of declining U.S. unconventional oil production in the period 5-8 years from now that has IEA's chief economist, Fatih Birol, scared. He has stated in various forums his concern about the current low prices inducing an unwarranted complacency about long-term oil supplies. Low prices have already induced a major tumble in oil rig deployments in the U.S., as we saw above, setting us up for a big dip in oil production.

We are also seeing major disruptions in the Middle East, in Iraq specifically, that increase concerns about global supplies when the U.S. fracking revolution runs out of steam. The UK's Telegraph paper stated in late 2014: "Mr Birol said current low prices presented a 'major challenge' for the relatively high-cost production of shale oil in the US, with investment expected to fall 10% next year." The IEA's 2014 World Energy Outlook, released in November 2014, warns about these long-term dynamics:

> The global energy system is in danger of falling short of the hopes and expectations placed upon it. Turmoil in parts of the Middle East - which remains the only large source of low cost oil - has rarely been greater since the oil shocks of the 1970s. Conflict between Russia and Ukraine has reignited concerns about gas security....
>
> The complexity and capital-intensity of developing Brazilian deepwater fields, the difficulty of replicating the US tight oil experience at scale outside North America, unresolved questions over the outlook for growth in Canadian oil sands output, the sanctions that restrict Russian access to technologies and capital markets and - above all - the political and security challenges in Iraq could all contribute to a shortfall in investment below the levels required. The situation in the Middle East is a major concern given steadily increasing reliance on this region for oil production growth, especially for Asian countries that are set to import two out of every three barrels of crude traded internationally by 2040.

There's also a possibility, however, captured in the EIA "high oil and gas resource" case in the 2014 Annual Energy Outlook, that U.S.

oil production will continue to climb through the 2020s and 2030s, peaking around 13 mbpd. This increase in production, if it occurs, will surely help exert a long-term downward pressure on oil prices, but given the very high decline rates of fracked wells it seems more likely that we'll see U.S. and global demand decline substantially in this timeframe, exerting the same downward pressure on prices. It is this idea—a peak in global oil demand—that I turn to next.

But what about peak demand?

The peak oil debate has taken an interesting turn in recent years in that it seems now that may hit hit peak demand before we hit peak oil (in terms of physical production limits). By peak demand I mean the globe may well see a permanent decline in oil demand before we hit long-term physical production limits. If this is the case the cornucopians will have been proven right rather than the peak oilers. To this I offer my well-worn phrase: time will tell.

Anyway, back to peak demand—why would we see a decline in demand any time soon, with the developing world population and consumption still growing strongly? Well, it's the same dynamic we've seen in earlier chapters on electric vehicles and other means of transportation.

We can't at this point make any firm predictions about sales of EVs even in the next few years, but we are seeing a classic learning curve with EV technologies that suggests a coming boom in EV sales. The real game changers will be the 200+ mile range EVs already discussed, including Tesla's planned Model 3, the Chevy Bolt, the new and improved Nissan Leaf, VW's planned 300+ range car, and probably many other OEMs already plotting developments in this space. Once these cars are available and affordable, we may well see the exponential growth curve turn sharply upward. Global plug-in electric vehicles sales exceeded 700,000 by the end of 2014, with

320,000 sold in 2014 alone. January 2015's global EV sales figure was twice the 2014 January sales figure (the latest figures available at the time of writing), so it appears that globally we're seeing a very robust sales growth trend continuing and we passed the 1 million mark in 2015.

Independently of EVs, we're also seeing many other technologies leading to lighter and more efficient cars more generally, regardless of whether the cars are electric or gasoline or diesel, as we saw in a previous chapter. Some kind of fuel economy standard is in place now for 80% of the world's passenger vehicle sold in 2013, according to the International Council on Clean Transportation. Figure 9 shows the vehicle fuel efficiency standards for the world's largest economies at the end of 2014. The general trend is toward cutting in half the emissions from passenger vehicles between 2000 and 2025. We're about halfway to achieving that goal and it is realistic to expect that the 2020-2025 goals will be met due to the many new technologies arriving that will improve fuel efficiency and decrease GHG emissions.

Figure 9. *Vehicle fuel efficiency standards for various countries (source: ICCT).*

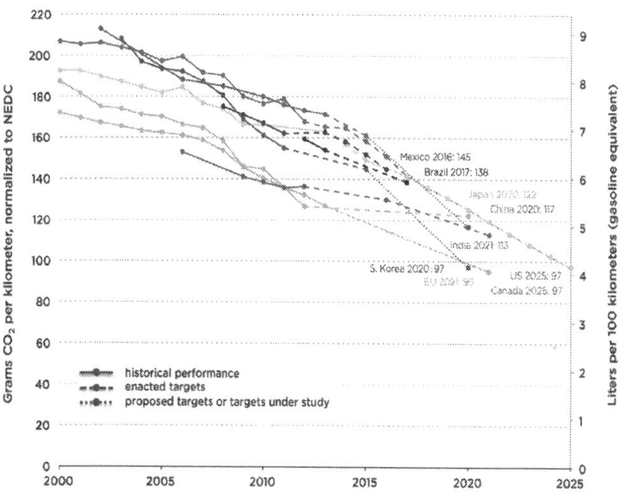

As a consequence of these trends (and many others), IEA projects in its 2015 World Energy Outlook that oil demand will rise to 104 mbpd by 2040, up from 91 mbpd in 2014. This is a significant increase but it's a large decline from previous forecasts that reflects the increasing role of improved vehicle fuel efficiency around the world. We can anticipate that we'll see additional vehicle fuel efficiency materialize in the coming years. We also have forecasts from China's oil companies that suggest that China's oil demand will peak far sooner than the IEA and others currently project. An early 2015 forecast from Sinopec projects that gasoline and diesel demand will peak by about 2025 in China. If China's development curve is mirrored by other major emerging economies like India and Brazil, we may well see peak demand before peak oil. An encouraging development that supports a relatively early peak in China's oil demand growth is the fact that 2014 saw a 220% increase in EV sales in China. China has shown with very serious growth in solar, wind, hydro and nuclear, along with coal power plants, that is capable of ramping up deployment of new technologies very quickly indeed.

Global oil demand depends on far too many factors to make any confident predictions at this point, however. But we can at least see a plausible future in which global oil demand peaks in the next decade due to major improvements in petroleum alternatives and due to ongoing conservation and efficiency efforts. And if this does happen we may actually be able to avoid the disruptions that would come from cyclical oil price spikes from inadequate supplies. So let's all cross our fingers that global oil demand will indeed peak before production peaks.

Chapter 17

A fun example of what could be: a fully solarized Burning Man

Burning Man is a crazy desert art festival and party in northern Nevada. It takes place on federal public lands and attracts about 60,000 people each year. People think nothing of driving ten or fifteen hours to get there, or flying in from the East Coast or Europe. It's an art festival that is radically inclusive and radically expressive. It's a good time.

I've been to Burning Man six times. I fell in love with the culture after my first time in 2008. I even have a tattoo to prove my love. (No, I didn't get the tattoo at Burning Man).

As amazing as Burning Man is in so many ways, it's not remotely green or sustainable. People generally bring diesel generators for power, or use built-in RV generators for camp power. And did I mention the driving and flying to get there?

There are many efforts, however, to green up the festival and we are starting to see an increasing number of solar panels in camps and art projects around the Playa. How can we leverage these early attempts at going green to fully solarize the Playa?

Snow Koan Solar is a long-time theme camp that provides solar charging for cell phones, computers, etc. They also provide (surprise) snow cones. All of this is, of course, free because everything is provided by camps free of charge at Burning Man, which is why it's called a "gift economy." People feel good based in part on how much cool shift they can give away! Snow Koan also provides technical support for those looking to go solar on the Playa. Snow Koan provides solar for its own village, about 300 people now, and they are looking at ways to expand their solar power to about 1,000 in coming years.

Black Rock Solar (BRS), a nonprofit entity affiliated with the Burning Man organization, offers solar panels on a rental basis to art installations, at $50 per panel. This is for the panel only, however, and does not include other components like wiring, charge controller or batteries; nor does it include installation. So this is a very DIY solution that is great, but necessarily limited in its impact.

Another option that has a lot of promise is for BRS or some other entity to offer community solar systems in each area of Burning Man. The Burning Man grid is laid out in a horseshoe each year, with streets named and marked. Each sector could enjoy the benefits of a large solar system with numerous outlets and extension cords running to the camps in that part of the city.

Who pays for this? Good question. Solar is getting way cheaper all the time, but it still costs a pretty penny for larger solar systems. For example, a 100 kilowatt system mounted on a rack on the Playa would cost about $300,000 today, fully installed. That ain't cheap by most people's standards. And if we add battery storage to ensure night-time availability of solar power it gets even more pricey.

A fee-based model seems the most promising. Under this model, BRS or some other entity would charge each camp that wanted to use solar power a set fee (say, $1000) per year. If each array powered fifty camps, BRS would receive $50,000 per year for one week of use. The other 50 or so weeks in the year would allow other use of the solar panels. While this model is not a financial slam dunk it does seem to hold some real promise.

Another option would be for each camp to pay for its own smaller system. For example, a 1 kilowatt system (four solar panels) plus batteries would provide enough power for most small camp needs. A "plug and play" 1 kW solar system is now available for about $3,000 online. This doesn't include inverter, charge controller or batteries. A battery backup system would cost another $1,000 or so, depending on the type of batteries. Small inverter, charge controller and wiring would bring the total to about $5,000. This, again, is not chump change. But each camp could try crowdfunding or seek community grants, or simply have camp members chip in for this long-term power solution as an investment in their future. Larger camps could scale up as required.

An intriguing new design that incorporates the inverter and batteries directly into the solar panels will soon be available. The Solar Liberator will soon deliver its first products after a highly successful crowdfunding campaign and time will tell how durable these products will be. The 500 watt option would be perfect for camps of 3-4 people and they're entirely modular so they can be scaled up easily for larger camps. You could literally place one of these panels on a car roof and plug in your extension cord for 24/7 solar power. This new technology would make gradual solarization of Burning Man much easier because of the hassle-free nature of this product.

The cool thing is that these small systems could be used at someone's home or business the other 50 or 51 weeks of the year. Under this model, state rebates and federal tax credits could also be used to significantly reduce the cost. The federal tax credit alone is worth 30% of the cost of the system (applicable only to the 500 kW Solar Liberator model).

Better yet, perhaps a SunPower or a Yingli Solar could be convinced to donate panels to a number of camps for one week, in exchange for some goodwill elsewhere. Burning Man is strictly a non-commercial environment, so no company logos or any kind of advertising are permitted. (You can't buy anything except ice and tea at Center Camp, but you can arrange for transactions like renting solar panels from BRS or renting a bike prior to arriving). This is part of what makes the Burn so special. But the event's noncommerciality does make it more challenging to engage in traditional approaches involving donated products in return for advertising or goodwill. In this case, the solar companies donating panels could certainly get the recognition that they deserve at fundraising events by camps prior to Burning Man.

The BRS solar panels available for rent were donated so there is potential to obtain far larger numbers of donated panels.

Another idea: BRS has donated dozens of systems to area nonprofits and schools after being used on the Playa each year. Perhaps these owners would allow BRS to borrow these systems back for one week each year to use on the Playa? This would be far cheaper than buying new systems since it would entail only the labor of deconstructing the systems from where they are currently and reinstalling them on the Playa for a week. The labor for deconstruction and re-construction is not insignificant, but, again, it's much cheaper than buying a whole new system with installation.

A final way to make Burning Man more green is through the purchase of carbon offsets. The carbon footprint of driving or flying to Burning Man is considerable. I'm not a big fan of offsets in general because I'm not convinced they're the best way to promote renewables or alternative transportation. That said, certified carbon offsets are available that are probably better than doing nothing about the issue!

In sum, there are numerous ways that the Playa could be solarized. The combination of individual and camp efforts with larger efforts led by BRS and Snow Koan could in just a few years make diesel generators a thing of the past. Lord knows there's enough sun on the Playa to run Black Rock City on it.

Chapter 18

Where the rubber hits the road: the role of communities in the energy transition

How can communities take control of their power mix? Who should own and control power generation assets in the 21st Century? In the U.S. this has been a long debate and today's landscape is dominated by investor-owned utilities, which are private entities that serve customers in order to generate a profit for shareholders. This model has worked pretty well for decades now but it's far from a perfect model.

In the U.S. today there are also many publicly-owned utilities (these are government agencies) and co-operatives, which are private but not operated to achieve a profit for shareholders, as is the case for investor-owned utilities. Co-ops are designed to provide power to areas where the investor-owned utilities dare not tread because it's not clear that the effort would be profitable.

There is a new kid in town, however, known as Community Choice Aggregation (CCA). CCA is a middle ground between full private utility control and full public control of power. Rather than controlling the entire power grid, like is the case in Los Angeles or Sacramento, the two biggest public utilities in California, CCA allows a city or a county to choose what type of power their citizens get and from where. The private utility continues to control the power lines and bill customers, so utility profit potential is preserved - but the control over power choices shifts to the CCA entity known as a "Community Choice Aggregator."

This middle-ground solution can provide significant local control over the power mix while avoiding extremely lengthy legal battles that usually follow local efforts to create a public utility like in LA or Sacramento (which took over twenty years of lawsuits before a resolution was reached).

Local governments are receiving more and more pressure from constituents to go green. With solar power prices plummeting steadily, the pressure to increase communities' share of green power is surely going to increase in the coming years. CCA presents a powerful tool for communities to be proactive in going green and providing more options to customers. By law, CCAs must meet procure at least as much renewable energy as investor-owned utilities but in practice so far all CCAs have used CCA to pursue even higher levels of renewables.

California's CCA law was passed way back in 2001 (AB 117) and it took almost a decade before the first CCA began operations. Marin County is the first CCA to get up and running in California, as "Marin Clean Energy." MCE began operations in 2010, so there is now a track record to look back upon. MCE offers customers three options for power choices: either the default 50 percent renewable energy mix, a 100 percent renewable "deep green" mix for a small premium (about $5 a month), or a 100 percent local solar option that began in 2014. The deep green mix is based on wind power renewable energy certificates.

Sonoma County was the second CCA off the start line and it has been operating since 2013 as "Sonoma Clean Power." Similar to MCE, SCP offers two options for green power: the CleanStart option for 33 percent renewable energy at a cost savings compared to PG&E and the EverGreen option for 100 percent renewables at a small premium. SCP also offers right on its homepage a convenient button for opting out of the CCA entirely.

Both MCE and SCP offer a local feed-in tariff option for wholesale projects up to one megawatt. SCP's option is called ProFIT and offers . MCE's option is called (surprise) its Feed-in Tariff program. Neither of these programs have seen many projects installed yet, so time will tell if these programs are effective in generating local power.

The City of Lancaster, north of Los Angeles, voted to explore CCA in 2014 and is slated to begin operations in 2015 as "Lancaster Choice Energy." If it's successful it will be the first CCA in Southern California Edison territory (the previous CCAs are in PG&E territory). It expects to offer rates 3 percent below Edison rates and its default power option will include 35 percent renewable energy. As with Sonoma and Marin customers can also opt for a 100 percent renewable energy mix for a premium.

The Santa Barbara region is also exploring CCA at this time. There was some interest a few years ago but this faded until a new push by the Community Environmental Council, a local environmental group, and other advocates in the region renewed their efforts. The City and County of Santa Barbara plus some other jurisdictions are exploring the feasibility of CCA in our region.

Various other communities around the state are also looking at CCA, including San Francisco, a group of cities and the county of San Luis Obispo in the Central Coast region, another group in the Monterey area, and a group of cities in the South Bay area of Los Angeles.

Marin County's partnership options

The truly exciting development with CCA at this point is Marin County's expansion to jurisdictions far from Marin County. There is nothing in the law preventing cities and counties from all over the state joining forces in a single CCA. This is the case because CCA is a financial tool and doesn't require contiguity or even that the partner jurisdictions are in the same utility territory.

MCE expanded in 2013 to include the city of Richmond, in Contra Costa County. Service to the unincorporated areas of Napa County began last month and the cities of Benicia, El Cerrito and San Pablo, will begin in May of this year.

MCE is thinking even bigger, however, and is open to partners that aren't in counties adjacent to Marin County (as all of their partners have been so far) as "Special Consideration" members. MCE has offered to complete feasibility studies for potential new partners at about $25,000 each. This is a relatively small amount considering the potential savings and other benefits that may come from CCA.

The importance of "opt out"

A key feature of the CCA law is its "opt-out" language. This means that if a city or a county chooses to become a CCA, the residents of that city or county are enrolled automatically in the CCA unless they choose to opt out. They're given many opportunities to do so and the law requires notification in various ways of the ability to opt-out, so there is nothing coercive about this process. Community Choice really is about choice.

The road has been rocky at times for CCA. The law has weathered at least two major challenges so far: a ballot initiative (Prop 16) pushed by PG&E in 2010, with $46 million of PG&E shareholders' money behind it and, in 2014, a legislative challenge in Sacramento, AB 2145. AB 2145 would have effectively gutted the CCA law by requiring CCAs to follow an "opt in" rule rather than the "opt out" rule described above. Based on inertia alone it's generally difficult to induce customers to switch utilities so requiring opt in rather than opt out would for all intents and purposes have killed CCA. Luckily, this bill died in late 2014 when it wasn't called up for a final vote in the state Senate.

CCAs can be formed by a vote of the city or county council, or by a vote of the people. My preference is to have the people vote because I'm as a matter of principle a strong small "d" democrat: more democracy is almost always better than less democracy. I can recognize, however, that there might be circumstances where a city or county council vote may be better strategically.

Community Choice is about choice, local empowerment, and higher renewable energy goals. As we've seen so far with the track record of CCA around the state, these goals can be met while also allowing some customer savings, and that's pretty remarkable. It's too early to call CCA an unequivocal success in California, but the next few years will give us enough experience to know whether or not CCA lives up to its promise.

Chapter 19

Even closer to the ground: taking your home off-grid

Solar energy has been a passion of mine for well over a decade now, but it wasn't until recently that I finally was able to install my own solar panels. I've owned a couple of condos in Santa Barbara over the years, where I normally live, but the location of each condo and my small power bill ($20-40 per month) never justified going solar. I've now been renting in Santa Barbara for a couple of years and have been urging my landlord, to no avail, to go solar.

I finally got to go solar when I bought some land in early 2014 on the Big Island of Hawaii that included a run-down cabin. I decided to invest some money to retrofit the cabin (Figure 1) and convert it into a liveable tiny home slash writer's cabin slash vacation rental. And I decided to go entirely off-grid, including for power.

Water falls from the sky and is stored in a catchment tank. With filters and a water pump, this is a sustainable and healthy solution for water that doesn't involve the authorities at all.

Figure 1. *My tiny home near Hilo, Hawaii.*

Hawaii has been making news due to its high saturation of solar power on the grid. Oahu, the most populous island, and the Big Island, the biggest island, have well over 10% of homes opting for solar. But these figures include only grid-tied homes. Hawaii's utilities have dramatically slowed interconnection of new solar facilities due to increasing concerns about the stability of the grid. Going entirely off-grid is one solution to this growing problem, which has many other benefits, but also some downsides that I'll discuss here.

My little cabin is up a fairly long driveway and it would have cost many thousands of dollars to have my local utility, Hawaii Electric Light Co. (HELCO), install power at my place. This made going off-grid an easy decision. Since my cabin is small and the climate is warm all year around, it didn't need a very big solar system. As preparation, I read Michael Boxwell's *Solar Electricity Handbook, 2014 Edition* and with some help from a local solar installer who specializes in off-grid systems I figured out the appropriate size and equipment I'd need.

I wanted to do as much of the design and installation myself because I've been wanting to learn more about the nitty gritty of solar for some time. This was a perfect opportunity to get my lawyer hands dirty and take my eyes away from staring at my computer screen for a while.

Boxwell's book is a great resource and even though he's a Brit the advice applied just fine to the U.S.. I decided to build a 24 volt system. Combining four batteries serially got me to 24 volts. Ditto with two solar panels, which are 12 volts each.

In looking at my power needs, I calculated that I could probably get by with just two 250 watt panels, for a total of half a kilowatt. My equipment list included, accordingly: the two panels (Schott Solar, bought from Provision Solar, a Hilo-based company owned by Marco Mangelsdorf, costing about $800 including racks to install them on my roof), a ProStar charge controller (bought from Off-Grid Solar Specialists for $350), a Schneider Electric inverter (from the same store, $1,200), four Interstate batteries ($130 each, bought from Pahoa Propane and Batteries), and a bunch of wire and battery connectors.

I did most of the installation myself, painstakingly and with much trial and error. I ended up needing an hour of a professional installer's time to wrap it up, so the total cost of my system was about $3,500, or $7 per watt installed.

The batteries are 232 amp-hours each, but because they're connected in serial fashion, they supply a total of 232 amp-hours at 24 volts. They are charged directly from the charge controller at 24-30 volts. The inverter connects to the batteries and the home's central power panel connects to the inverter. When I leave for extended periods I turn off the inverter but leave the charge controller running. This allows the panels to keep the batteries fully charged and healthy, while avoiding the risk of something going awry with the home power that could leave my batteries dead again.

The ProStar charge controller is addictive to watch since it shows how much power is coming in from the panels and how full the batteries are. I think of this as “watching TV.” I’ve seen as high as 38 volts coming in during very sunny times at mid-day. I was also pretty happy to see how much power comes in even when it’s cloudy or rainy—from $1/10^{th}$ to $1/4^{th}$ of full power. Power production starts as soon as it gets light but doesn’t really get going until a few hours after sunrise.

Figure 2. *ProStar charge controller at my house.*

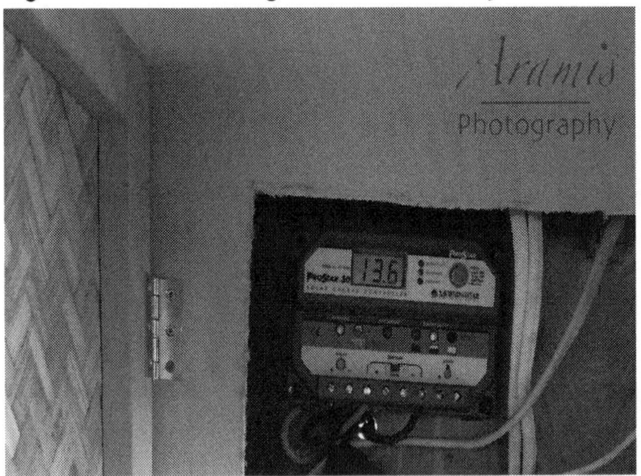

My pre-installation calculations showed that my little system could power a small fridge, lights and a TV and DVR player, as well as chargers for computer and cell phone. I was happy to find that my calculations were accurate, but I did get worried a few times when I woke up in the morning and found my batteries very low, as evidenced by the blinking red light on the charge controller. But invariably by noon or mid-afternoon my batteries go back to green.

My new lava rock wall shower, fed by rainwater and powered by the sun, was in business.

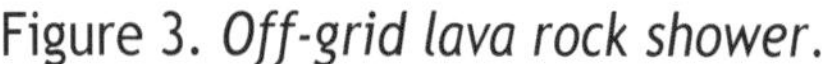

Figure 3. *Off-grid lava rock shower.*

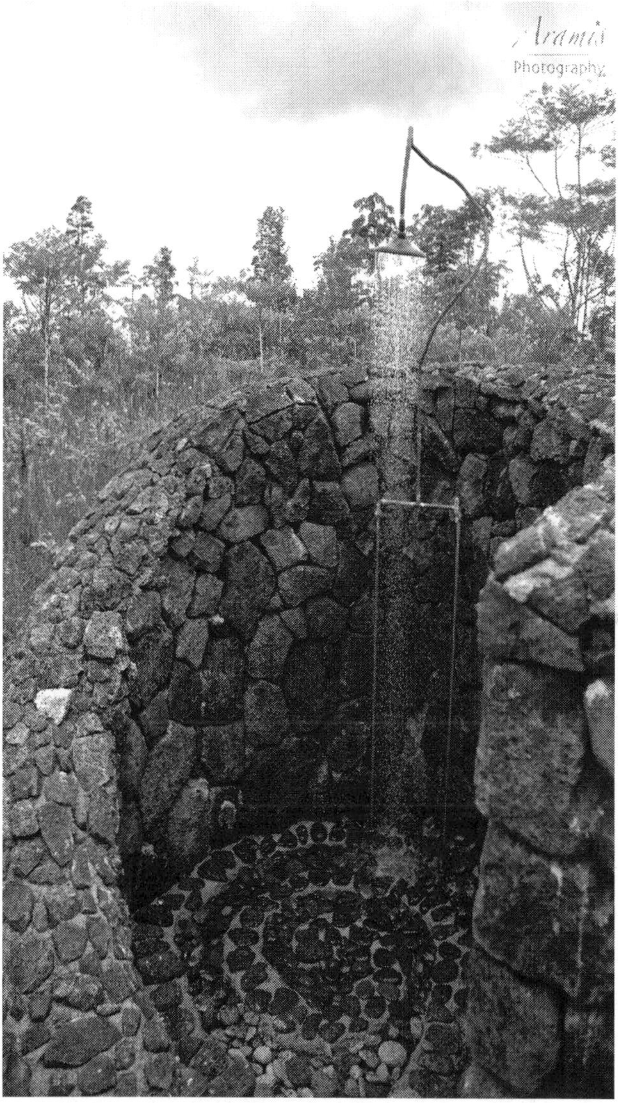

It can be challenging at times being off-grid, particularly when you're working out the kinks. I initially installed the system in April this year and I returned to Santa Barbara for a few months. When I went back to Hawaii in July I found my solar system down and my batteries dead dead dead. I couldn't figure out what happened so I called the same local installer who had helped me wrap up installation in April. He diagnosed loose battery connections as my problem. I had to take the batteries back to the store to charge the batteries back up. That took three days, for various reasons, so I was living without solar power for a week.

I didn't mind it too much since I have plenty of candles and also bought a kerosene lantern. I kind of liked the primitive living conditions and I have a backup generator for when I needed to charge my computer and phone. I'm also rarely at home at night so it wasn't a big deal. The fridge was the only thing that I really missed. But since I was only in Hawaii for three weeks on that trip I wanted to get the system back up and running so that I could rent out my place while I was gone.

I was able to get the batteries re-installed and wired in snugly, but still my system refused to come back to life. I double checked everything and called the installer again. No help. I went back to the store where I bought the charge controller and he agreed to give me a new one since it seemed that the problem must have been the charge controller. I re-installed it. Still nothing. Long story short, we finally figured out that I had been given the wrong configuration to install my panels and the panels were in fact supplying twice the voltage that the charge controller could handle. So it had burned out. I re-wired the panels and finally the whole system snapped back to life. It's now been running fine for over a month and it should keep running fine for a long time (fingers crossed).

Here's where it gets interesting. Another major benefit of being off-grid was highlighted with the recent hurricanes that threatened

Hawaii. After Iselle hit the Big Island last Friday, downgraded to tropical storm status, well over 20,000 utility customers went without power and are still without power at the time of this writing (Sunday). Here's a picture of the power line on my road:

Figure 4. *The power line on my road after Tropical Storm Iselle.*

My tiny house, on the other hand, is sitting pretty with solar power and batteries going steady.

In sum, going off-grid in Hawaii can be very liberating and can save money too. Power bills are high in Hawaii since most power is generated by imported diesel. My power bill would probably be about $100 a month if I was on-grid, but I'd also have to pay the large fee for building a line to connect to the grid. With my off-grid system all of my costs are up-front and there are no ongoing costs except when/if things break down or go awry (again, fingers crossed).

I can't calculate the payback time accurately since I don't have a power bill to compare. Based on my estimated $100 monthly bill I would, if I had been on-grid previously, pay off my system in about three years; but that again leaves out the cost of connecting to the grid in the first place.

Going solar also allows me to avoid the emissions of dirty diesel, to avoid spikes in electricity rates, to avoid dealing with the utility on interconnection or net metering, and to be secure when the grid is down due to storms or falling trees.

I'm still very new to off-grid solar and I may find new issues cropping up. I'll report back if anything newsworthy happens. My tiny home in Hawaii is available for vacation rentals most of the year, so you too can witness the benefits and challenges of off-grid solar firsthand.

Chapter 20

The cost of the future: what will the transition cost?

What follows is an email dialogue yours truly and David Victor, professor of international relations and director of the Laboratory on International Law and Regulation and co-author of a recent U.N. report on the cost of mitigating climate change. A piece in the *New York Times* by Andy Revkin prompted me to initiate this dialogue. David agreed to publication here.

Tam: I was surprised to see your comments on the cost of carbon mitigation in relation to renewables in the *New York Times*. My view, supported by experience and a lot of good studies about the current and future costs of solar and wind, is that renewables are already cost-savers in many contexts and will increasingly improve in terms of cost savings vs. fossil fuel energy sources. I'm curious regarding what data you use to support your projections for high costs for renewables when compared to fossil fuel sources? And what geographic scope are you considering in your analysis?

David: Thanks for your note, Tam. I am thinking globally and attentive to renewables at scale. In some special circumstances -- usually with very expensive rival power and very good physical conditions for renewables, such as in Hawaii where rival power is costly oil and there is a lot of sun -- renewables can scale a bit on their own. But the vast majority of modern renewables don't scale on their own without massive policy support, including grid integration rules that hide the full cost. That doesn't mean we shouldn't "do" renewables -- that's not what I am saying. But what it does mean is that renewables are still being improved and they are far from ready for scale applications, and when you are talking about cutting emissions it is scale that matters.

I am mindful that there are various studies making (wild in my view) claims about the ease of quickly (i.e., a few decades) shifting almost fully to renewables. But reality is setting in, even for places in the world that are committed to this vision. Look at Hawaii where the power company has had to suspend new solar installations because of grid integration issues. Look at Germany where they are re-installing a massive investment in inverters because most of the old installations could not integrate at scale into the grid. Worse, in Germany even Angela Merkel -- a huge renewables fan -- has signaled that the current feed-in tariff (FIT) policy is totally unsustainable. Indeed, the latest Energy/Environment White paper from the EU with targets for 2030 has set a soft 27 percent goal for the EU for renewables and adamantly refused to allocate that goal to individual nations -- which would make it enforceable -- because the EU has no clue how it can reach those high levels.

The problem is scale, and it doesn't mean that the industry can't be competitive in special markets where customers are wealthy and nobody really notices the cost -- California is an example, where an aggressive RPS along with other investment policies drive investment in renewables that probably wouldn't happen [otherwise].

Tam: While I certainly agree that getting to high penetration of renewables around the world won't be easy, I don't think it will be costly. Rather, I think it will be a substantial cost-saving opportunity for the large majority of jurisdictions that do it. You mention Hawaii and Germany, so I'll focus on those in my response as examples of this transition. I'm actually living in Hawaii now (I split my time between Santa Barbara, California, and Hawaii), so I've followed Hawaii's trajectory pretty closely.

HECO and HELCO, the utilities on Oahu and the Big Island, respectively, are indeed claiming issues with high penetration of net-metered solar, and they slowed down approvals for new net-metered solar projects in 2013 and 2014 pretty substantially. This highlights the remarkable growth rate in solar in Hawaii, which has resulted in some of the highest solar penetrations in any jurisdiction in the world.

However, the utility slowdown in processing interconnection applications will be a temporary delay, and this is a technical issue, not a cost issue per se. The policymakers and the public are fully behind Hawaii's solar tax credit program and the retail price credit that net-metered systems receive for excess power that is sent to the grid. Moreover, and more importantly for this discussion, Hawaii's other programs for renewables, its renewable portfolio standard wholesale procurement program and its new feed-in tariff program, which is also focused on wholesale procurement, have been found recently to be highly cost-effective for ratepayers. A recent report by E3 for the Hawaii PUC found the following: "We find that renewable energy provides a significant opportunity for Hawaii to reduce electricity costs to customers. There are many renewable technology types that provide net value to ratepayers. These include various sizes of wind energy and solar photovoltaic generation on each island, as well as in-line hydroelectric generation. Given the high costs of purchasing petroleum fuels for energy on the islands, these approaches can lower utility costs."

I've been told personally by HELCO representatives that they view renewables as a substantial cost-saving opportunity for ratepayers, and they are looking to procure new renewables at far less than avoided cost of diesel-powered generation (which is well above $200 per megawatt-hour and rising). The islands already have substantial backup capacity in the form of existing diesel-powered stations, so balancing variable renewables is not really a technical issue in Hawaii. The issue is how to deal with excess power backflowing from distribution circuits. But there are many technologies, including smart inverters and battery storage, that can and will solve these issues in the coming years.

Based on the E3 and other analyses, I am currently working to build a coalition here in Hawaii to accelerate the renewables transition and to get the Big Island in particular to carbon neutrality by 2030 or sooner.

Turning to Germany, again I agree with you that there are some temporary technical issues that they are facing from their rapid buildout of renewables. However, the proof is in the pudding: Germany has transformed itself from almost no renewables twenty years ago to over 25 percent of its electricity coming from wind, biomass, solar and hydro in 2013 and higher in 2014. This is remarkable. Not only has Germany transformed the global renewables market (particularly solar) by spurring huge reductions in the cost of equipment, the country has transformed its own market to the point where solar is now a net cost saver for almost all customers. A recent study found that solar is very close to being cheaper than coal EU-wide -- and credit for this remarkable development can be laid at the feet of far-sighted German policymakers.

This study found that solar power already costs as little as 8 euro cents per kilowatt-hour and will likely fall to about 6 euro cents in coming years. Wind power is already 5 to 11 euro cents/per kilowatt-hour. Current costs for coal and natural gas range from 5 to 10 cents per kilowatt-hour, and these costs are surely going to increase. So the economic benefits of renewables are already apparent, and they grow increasingly positive over time. Even when we add integration costs, in terms of backup power to balance variable renewables, and new transmission as required, renewables still come out looking very good on cost alone.

I was surprised to see the EU's watering down of country-specific renewables goals, but I am very optimistic that the EU will, as a whole, far exceed these goals based on market incentives alone. Natural gas is expensive in the EU, and coal isn't much cheaper. Wind, solar and biomass are looking very attractive to more and more countries based on economics alone, and this trend is set to explode in the next decade or two.

So, yes, there will be technical problems in reaching higher penetration of renewables in every jurisdiction. These problems will add to the cost of generation from renewables, but on balance, renewables will still [represent] a net cost savings in the large majority of jurisdictions around the world in the coming years. And this is the key concept to keep in mind: while renewables can be cost-effective even today (on an apples-to-apples basis, even accounting for subsidies), this cost-effectiveness is on a dramatic, long-term upward arc, because renewable energy costs are getting cheaper and cheaper, while fossil fuel costs are generally getting more and more expensive. Yes, natural gas in the U.S., as well as oil production, has experienced a recent renaissance, but this is a short-lived phenomenon. Natural gas costs are back at $6 (up from $2 just a couple of years ago) [since the time of writing, natural gas has fallen back to below $3], illustrating not only the wild volatility of fossil fuel costs but also their unreliability in terms of long-term planning.

Solar is booming, both in terms of installations and in terms of company stock valuations (SolarCity, First Solar, etc.). The TAN solar ETF is up 25 percent year to date. Solar installations surpassed 40 gigawatts in 2013, up from 28 gigawatts in 2012. 2014 promises to be bigger yet. Wind installations now exceed 300 gigawatts worldwide (solar is catching up and is now over 100 gigawatts) and still growing well, though slower than solar.

We are already seeing solar and wind at scale. So we can in fact sit back and enjoy the ride, because exponential growth trends in these industries are clear and will very likely continue in the coming decades. We're not out of the woods on climate change because the transportation industry is the tough nut to crack, but in terms of electricity, I'll happily wager that by 2030 half of the world's electricity will come from non-fossil sources.

The bottom line is that wind and solar are well on their way to growing very well without any government support in many jurisdictions around the world. And this is unequivocal good news.

David: Thanks for your note. I'd like to clarify three things.

First, I expect that Hawaii will be cost-effective for lots of renewables for one simple reason: thanks to oil-fired generators (which are typical when you have relatively small, island-based power networks), it has the most expensive electricity in the nation by a long shot. What happens in Hawaii in terms of relative competitiveness tells us basically nothing about the rest of the world. Given the high cost of incumbent electricity in Hawaii, you could generate electricity from starlets riding bicycles and it might be cost-effective. Sure, the buildout in Hawaii and Germany (and Denmark and a few other places) is remarkable. But the key questions revolve around whether these are typical places and whether the policies (notably in Germany) are sustainable at scale. We clearly have different points of view on this.

Second, I also suspect that when we look closely that the integration issues we are seeing in Hawaii (or California or parts of Germany or many other places) that these will turn out not just to be "temporary delays" for mere "technical issues." There are fundamental problems in managing a grid at very high reliability with large amounts of variable and intermittent power. Some of that might get addressed with incentives to build storage (as we are seeing to some degree in Texas) or mandates to build storage (as now unfolding in California and other places). But the fundamental properties of that grid are totally different from the "normal" grid. Add into that a large role for distributed energy resources -- such as rooftop solar or onsite self-generation at industrial sites -- and the challenges for grid management and planning are huge. This is why the Electric Power Research Institute's new "Integrated Grid" initiative is important, [as are] lots of complementary efforts. These are surmountable challenges, but they require lots of rethinking and planning and huge room for error if rushed.

Third, I have not done the detailed spadework that is needed on the Fraunhofer study that you linked to, but I've seen a lot of these studies over the years, and I'd urge all of us to look closely at the key assumptions that drive the outputs. Usually, the most important assumptions are a) the assumed cost of capital and financing structure; b) the assumed cost of fuel; and c) the assumed costs of integration. Very quickly, look at table 2 (page 11 in the linked study) and you'll see what drives the analysis, which is the combination of very low financing assumptions for renewables (and high assumptions for fossil plants). Those aren't real, market numbers -- they must be a fiction that reflects other policy incentives at work. Does anyone really believe that the market by itself would finance small PV with an 80/20 debt/equity ratio where the acceptable risk-adjusted return on equity is 6 percent and debt pays only 4 percent, while radically different financing assumptions are used for central power stations? And then look at the operation costs -- notably high numbers for brown coal and even for gas. I am not arguing in favor of brown coal -- quite the opposite, as I think it is bizarre that an environmental leader still burns brown coal, but such is the power of the coal unions -- but [rather pointing out] that we need real apples-to-apples comparisons.

A few more wrinkles to the analysis just to make it clear how problematic the case will be. The out-year assumptions on gas prices are really high (see table 5 on p. 15), which is probably hard to sustain if you think the rest of the world is in the midst of a gas revolution that will (as in the U.S.) bring down prices. But those assumptions make fossil fuels look unattractive. And the Fraunhofer study, as far as I can tell, hasn't yet seriously reflected the grid integration costs -- which is hardly surprising since everyone in the analyst community is still trying to get their heads around that question. (Chapter 6 of the Fraunhofer study basically outlines some long-term visions for how that might unfold, rather than actual analysis.) Again, I am not criticizing the Fraunhofer study -- in fact, they do some of the best work on this topic in Europe -- but simply drawing our attention to the kinds of assumptions that drive analyses and raising serious questions about whether those are scalable.

Indeed, I suspect it is exactly those kinds of concerns that help explain why the new EU white paper envisions massive cuts in emissions, massive expansion in renewables, and big reductions in power costs all simultaneously -- without a clear vision for how that will be implemented in reality. The reason is that so much of the work done on competitiveness of existing renewables doesn't grapple with grid integration seriously and does somewhat simplistic levelized cost of electricity (LCOE) calculations within power markets, like Hawaii, where the fossil incumbent is terrifically expensive.

Tam: You raise a number of additional points and I'll address them one by one.

1. You argue that success with renewables in Hawaii is irrelevant to the broader issue of scaling of renewables. Scaling of particular technologies will, I agree, be necessary for renewables to be a big part of the solution to climate change. I discussed Hawaii because you raised Hawaii in your previous email as an example of problems with integrating high levels of renewables. The integration issue is

why Hawaii is very relevant to the scaling discussion. Yes, Hawaii's electricity rates are very high, helping to make renewable energy more competitive than in other jurisdictions. But Hawaii is actually very relevant because it is a laboratory for showing how jurisdictions around the world can deal with high penetration. We can both look forward to watching Hawaii's integration efforts unfold in the coming years.

On the cost issue, Hawaii's cost differential between high-cost diesel power and low-cost renewables will in fact be mirrored increasingly around the world because of the two major background trends that have become quite clear in recent years: 1) increasingly low-cost renewables as they reach scale (solar panels, for example, have come down in cost over 50 percent in the last few years alone); 2) increasingly high-cost fossil fuels. Oil costs have stayed remarkably high even as the world struggled economically, which suggests that when the global economy recovers fully, we'll see far higher oil prices. As I mentioned in my last email, natural gas costs in the U.S. were quite low for a few years after the U.S. economic crisis, due to [recently developed] fracking techniques and lower demand, but we are now seeing costs shoot up again. Prices for natural gas are far higher in Europe and Asia. This increase in natural gas costs will make renewables increasingly cost-competitive in the large majority of jurisdictions where natural gas is prevalent in power generation.

2. You argue that integration of renewables at high penetrations will raise major long-term issues. Time will tell on this one, but numerous reports, cataloged by Lawrence Berkeley Lab and others, have found that the costs of integrating high penetrations of wind and solar are not actually that high. For example, the 2012 LBL annual wind power market report stated, with respect to U.S. markets: "Recent studies show that wind energy integration costs are below $12 per megawatt-hour -- and often below $5 per megawatt-hour -- for wind power capacity penetrations of up to or

even exceeding 40 percent of the peak load of the system in which the wind power is delivered." This is about a 10 percent premium over the cost of energy from wind, which is entirely affordable and does not represent a major economic or technical challenge. Even with these integration costs, wind power is highly competitive in most U.S. markets.

Germany again provides a good example with respect to its integration of wind and solar. This analysis, again from the Fraunhofer Institute, found that changing just one grid parameter in Germany would allow far higher penetration of renewables at lower cost. (Changing must-run thermal power plants from 25 gigawatts to 20 gigawatts, the technical lower capacity, allowed variable renewable penetration to go from 25 percent to 40 percent -- a remarkable change with one technical tweak to the system). My broader point is that we are already witnessing the renewables tsunami break (beneficially) over the world's power systems, and we are managing collectively very well in dealing with higher penetrations in those jurisdictions like Germany, Denmark, Hawaii, etc., where issues have arisen. Again, I'll happily wager that we'll see the world's electric power system 50 percent or more fossil-free by 2030.

As we reach half or more of our power coming from variable renewables, in the U.S. and globally, which will likely start to happen by 2030 or even sooner in many jurisdictions, battery storage and smart inverters (to handle reactive power issues) will become increasingly important. For this reason, I strongly support jurisdictions like Texas, California and Germany, which are incentivizing or mandating new energy storage procurement programs. I note also that California's new 1.3-gigawatt energy storage mandate is a cost-effective mandate, which means that it can't lead to any net cost for ratepayers.

3. Regarding the Fraunhofer report on solar and wind becoming cheaper than coal power in the EU, you argue that the financing cost assumptions are off for renewables vs. fossil fuels. I believe these are in fact accurate assumptions based on real-world costs, which stem from the feed-in tariff policy that Germany has had for over a decade for renewables. The feed-in tariff requires utilities to offer twenty-year contracts to renewables, at a set price. The low cost of money for renewables, expressed in the Fraunhofer report, is a key benefit of feed-in tariff policies. The predictability and low risk that the feed-in tariff creates, along with the low risk of solar and wind due to zero fuel costs, leads to much lower cost of financing. The same can't be said of fossil fuel power plants, which don't enjoy a feed-in tariff and of course have highly volatile fuel costs.

You also argue that the out-year fossil fuel cost assumptions in the Fraunhofer study are unrealistic. I'd argue, again, that we're already seeing a much higher price regime for fossil fuels in 2014. I think their cost projections are entirely realistic, and we may well see much higher prices. A major benefit of renewables is that we don't need to deal with such speculation, because there's zero fuel cost for wind or solar, and the prices paid for power can be known with certainty for literally twenty years or more. This is a major and unsung benefit of renewables.

***David*:** On gas prices, in a very cold winter, we always see price spikes, especially at trading hubs on the demand side of transportation bottlenecks. But the markets settle, and January/February is always a bad month to use as a base for making long-term predictions. Indeed, the markets themselves think gas will be settling around $4. (See NYMEX futures, for example, summarized monthly at: http://www.eia.gov/naturalgas/.)

Regarding integration costs, there are lots of studies -- some say costs are high, some low -- but my fear is that the studies that point to low integration costs don't rigorously look at these issues in the context of guaranteeing very high reliability. There's a world of difference between a power grid that works about as well as our internet connections (most of the time, except during periods of congestion) and one that is reliable almost all the time, as is required for modern grids.

Chapter 21

How to invest in our energy future

The renewable energy industry is well past its training wheel phase and there are now many ways to invest in all types of renewables. I've been an investor in renewables off and on for the last couple of decades and I offer in this chapter a little nonprofessional advice about how best to get into this field as an investor.

Deutsche Bank has become increasingly bullish on the potential for investing in the solar field, and they provide detailed research reports (for free) to back up their conclusions. An early 2015 report, which described their expectations for grid parity being achieved in 80% of the world by 2017, stated their key conclusion from an investment perspective: "We write this report because we think solar has now become an investable sector and over the next 5-10 years, we expect new business models to generate a significant amount of economic and shareholder value."

(Again, I am not a professional investment advisor and this chapter should not be considered legal or professional investment advice in any manner. Where I have a financial interest in my recommendations I indicate such).

As with all investments, the two key things to consider are your risk tolerance and your investment horizon. I'll start with the least risky investments with the longest time horizon and move toward more risky investments with shorter time horizons.

A friend of mine gave me his lifetime of accumulated investment wisdom recently: figure out the big waves that you want to ride in the long-term and position your surfboard accordingly. The renewables revolution is a very big wave arriving more or less now, and more and more people are lining up to catch this wave. I'm not a trader—I'm a long-term investor—and this chapter is written to help people who are looking to invest, not trade.

The sure thing

We are now at the point where individuals and companies can invest directly in renewable energy projects, even in small amounts. These are very low risk investments as long as there is a contract in place to either sell the power (for wholesale projects) or to net-meter the project (for behind the meter projects), which is the case for my recommendations below. The risk is low, particularly for solar PV projects, because not much can go wrong with solar PV once the facility is installed and operational.

A couple of low-risk ways to invest directly in renewables:

- JoinMosaic.com and some other sites like solarcity.com allow you to buy a piece of a real solar project that is fully vetted by the investment teams. Interest rates vary from 4.5 to about 6 percent over a 5-6 year obligation period. Mosaic is currently not accepting new investors but will resume accepting new investors soon. Disclosure: I've invested about $1,000 in a few different projects at www.JoinMosaic.com. So far I've realized the exact rate of interest I was promised and was able to easily reinvest my earnings back into real solar projects (until the site stopped new offerings).
- Renewable energy bonds. Warren Buffett's MidAmerican Energy company offered $1 billion in bonds, at 5.375%

interest, to finance about half of the cost of its huge 550 megawatt Topaz Solar Farm in San Luis Obispo County in central California. Green bonds now total over $9 billion and there will be an increasing number of similar offerings available for qualified investors.

A little riskier, but higher potential returns

The go-to investment strategy appropriate for most industries is to buy stock in companies in the industry. In the last couple of decades, this strategy has become lower risk due to the advent of exchange-traded funds (ETFs) in all fields. ETFs trade like individual stocks but they are in fact a basket of potentially dozens of individual stocks. A few opportunities in the renewable energy industry:

- TAN - the Guggenheim solar industry manufacturing ETF that tracks the MAC Global Solar Energy Index. This ETF is an investment in an increasingly solarized future, and with solar growing at a breakneck pace around the world and in the U.S.—the world installed almost 40 gigawatts of solar in 2013, with China leading the way with a whopping 12 gigawatts—this seems like a good bet to me. Disclosure: I own some TAN shares.
- PBW - the PowerShares Wilderhill Clean Energy Fund is a broad renewable energy industry ETF that tracks the Wilderhill Clean Energy Index. This is a general proxy for particular U.S.-traded renewable energy companies. I used to own some PBW shares but lost quite a bit of money after this ETF peaked in 2008 and then declined sharply. However, this ETF has bounced back and may well continue that growth.
- LIT - a lithium mining and production ETF that tracks the Solactive Global Lithium Index. LIT is designed to track the

global lithium industry, which is essential for batteries in electronics devices as well as most electric cars and stationary battery storage systems, both of which are set for growth. LIT is an investment in an electric car future. LIT is considerably down from its highs in 2011 but my feeling is that the EV revolution is set to takeoff in the coming years and we'll likely see lithium demand soar. Disclosure: I own some LIT shares.

Riskier still, but higher potential returns

For those with a higher risk and/or a longer time horizon, there are hundreds of individual stocks one can invest in. As with all individual stocks, companies' stock prices can swing wildly in short time spans, particularly in the relatively young field of renewable energy. The best way to temper risk while still investing directly in renewable energy companies is to buy shares in companies that do more than just renewables. For example:

- GE (GE) is a U.S. company with a large exposure to wind power. GE Wind, fully owned by GE, is the biggest maker of wind turbines in North America.
- Siemens (SI), a German company that has large investments in wind turbines and solar panels.
- GM is a large U.S. auto company that is pioneering electric vehicles, starting with the Chevy Volt. If you believe in an electric car future, GM may be a good place to invest.

Some individual stocks focused entirely on renewables or electric vehicles may offer strong returns in the future. This category is akin to recognizing the next Google, Amazon or Apple before everyone else does. A small sample of the many options available now:

- Tesla Motors (TSLA) has taken the car industry by storm, rising from plucky Silicon Valley startup to titan-to-be-reckoned-with in just a few short years. Tesla's market share is now well over $34 billion (it was only $21 billion when I first wrote this chapter), rivaling that of many of its century-old competitors. While its stock price seems to be overvalued compared to even its forward earnings (it currently has no net earnings), this is a play based on Tesla's enormous potential market share as its vehicles find new markets and become increasingly affordable—as well as a play on the amazing track record of its founder, Elon Musk. Disclosure: I own some Tesla shares.

- SolarCity (SCTY) is another Musk creation, focused on solar installations in the U.S. It has also come from nowhere and has become easily the largest solar installer in the U.S. in just a few years, with about 1/3 market share in residential installations in 2013. While SCTY may also be over-priced based on current and projected earnings (it also has no net earnings currently), this stock is also a matter of recognizing the huge future market for solar PV technologies, which have come down in cost dramatically in the last five years and are now increasingly mainstream. Disclosure: I own some SolarCity stock.
- Cree (CREE) is a U.S. company focused on energy efficient products like lightbulbs and light-emitting diodes for various applications. It's been on a tear since early 2012 and may continue this tear into the future as LEDs, which are far more efficient and durable than Compact Fluorescent Lightbulbs and other lighting technologies, take off.

Highest risk, but highest potential return

If you're entrepreneurial you can invest in your own renewable energy projects. Initial investment capital to create a saleable asset in the solar field, for example, is about $30,000 minimum, for a 1-3 megawatt project, which includes obtaining interconnection authorization, site control (own, option or lease) and a Power Purchase Agreement. With these items in hand you can often sell the asset for a handsome profit, though returns are being increasingly squeezed by the ever-higher number of participants in this space. This option is the focus of the next chapter.

In sum, there are a ton of opportunities to make money in the clean energy sector today. By doing your research and treading carefully, you can minimize your risks while also doing well.

Chapter 22

How to make money with solar on your land

We've seen, in terms that are abundantly clear, that solar energy is booming here in the U.S. and around the world. Solar has clearly turned a corner in recent years, transforming from exotic, expensive and do-goodery, to cost-effective, viable and just plain sensible in many circumstances. Policymakers in California and elsewhere have recognized this sea change in the cost of solar and there are now a number of programs available for landowners to sell solar power to their utility.

This chapter is a brief primer on how landowners can take advantage of the boom in solar to make money from their property, with a focus on California, still the biggest state for solar (by far; in 2014, California installed as much solar as the rest of the country combined).

The large majority of California has enough solar resources to make solar viable, though some coastal areas are dicey. Assuming that you have a good enough solar resource, here are the key steps in developing solar on your property. The basic strategy outlined here can also work for those seeking to buy or lease property specifically for solar project development.

1. Figure out how big a project you can fit on your land

A good rule of thumb for ground-based solar projects is 6 to 8 acres per megawatt. Rooftop systems (which don't have tracking systems and thus can take up less land) require about 4 acres per megawatt. A megawatt of solar provides enough power for about 200 homes and will cost about $2 million today as an "all-in" cost. For ground-mounted wholesale solar projects (which sell power to the utility), you want to focus on projects at least a megawatt or more in order to justify the development costs. Some key programs in California are now focused on projects that are 3 megawatts or less, so I'll assume for the rest of this chapter that the project being developed is 3 megawatts (about 20 acres), even though you may want to develop a larger project.

2. Interconnect your project

Interconnection is the first major hurdle for project development. Interconnection means you have permission from the utility to connect and operate your solar project in parallel with the utility grid. They can't say no to you, by law, but many areas are just too expensive to interconnect viably, so it's highly important to figure out early on whether your project can be interconnected affordably.

California offers a viable Fast Track process for interconnecting projects up to 5 megawatts far faster than under other options. In practice, however, Fast Track is generally only available for projects 3 megawatts and below.

A very good tool for scoping your project's interconnection potential at no cost is to look to your utility's online interconnection maps. PG&E, Southern California Edison and San Diego Gas & Electric (California's big three private utilities) all maintain their own interconnection maps. You can find what wires are on your property and what voltage they are from these maps. You can also find out how much capacity is available to interconnect on your property.

You won't, however, find reliable information on whether your project will pass Fast Track. For that, you should submit a $300 request for a Pre-Application Report (PAR) to the utility. This new option provides an additional level of detail about your potential project site, above what is available in the interconnection maps. You often can, with some analysis, figure out if your site is likely to qualify for Fast Track.

The only way to be sure you can qualify for Fast Track, however, is to apply for Fast Track and go through the process. And that will cost some money. The application fees are quite low--$800 for initial review and $2,400 for supplemental review—but the real costs come from engineers and consultants who are required to create engineering diagrams and to shepherd the applications through the process. Total costs for obtaining permission to interconnect under Fast Track are usually about $25,000 to $35,000.

Even though it's called Fast Track, it can still easily take six months for full interconnection approval and another six months for required upgrades to be constructed.

3. Obtain a power purchase agreement

Obtaining a power purchase agreement (PPA) is now the biggest hurdle to development in California and other states. There is a ton of competition for PPAs in limited programs, so it takes some real strategy to obtain a PPA. Each program is different so there aren't too many generalities I can share about obtaining a PPA. Nevertheless, here are a few insights about current programs.

ReMAT

ReMAT is a new program that started in late 2013. Pursuant to SB 32, a law passed a few years ago, California utilities are now accepting contracts under the ReMAT program for solar and other renewables at up to 3 megawatts in size (1.5 megawatts for SDG&E). The program size is very limited, however, with only about 100 megawatts for PG&E and SCE each and fewer for SDG&E. This means that competition will be fierce for PPAs. The prices also go up and down based on market interest so there's not really any way to have good visibility for the PPA price you can expect under this program.

RAM

The Renewable Auction Mechanism (RAM) is an auction program for all types of renewable energy projects 20 megawatts or smaller. Rather than an adjustable price, as in ReMAT, RAM pricing depends entirely on what developers bid into the program. And winning bid prices aren't revealed for three years, so it requires some art to determine the best price to bid in each round of RAM. This is a much bigger program than ReMAT, with about 600 megawatts of capacity still remaining through 2017. Projects smaller than 3 megawatts are generally not eligible to bid into RAM.

The new RAM program created in late 2014 and early 2015 includes the Green Tariff Shared Renewables program (SB 43) that allows utility customers to buy a share of actual projects. The projects, up to 20 megawatts, must be procured through RAM and the California Public Utilities Commission approved this 400 megawatts program in early 2015. This is a major boost to the RAM program and, along with the additional 275 megawatts (and perhaps much more) added in late 2014, has breathed new life into this program.

4. Permitting

Permitting is generally the easiest step in the development process, particularly for smaller projects like I'm focusing on here. At 3 megawatts, a solar project is about 20 acres and can in some counties be permitted with a mitigated negative declaration under CEQA. A full EIR may be required for larger projects, but by avoiding the EIR requirement, smaller projects can capture some negative economies of scale in terms of both lower costs and faster permitting times.

Once a developer/landowner fully entitles the planned project (interconnection, permitting and PPA), it can either be flipped for a profit or built out for long-term revenue. Financing of projects in today's markets can be extremely complex so the flip model is generally the default for developers who aren't large companies.

Return on investment varies widely. The costs of development and the price of power to be sold under the PPA are the primary causes for variation. However, based on today's market realities it is not unreasonable to expect returns of two or three times invested capital for a flipped project, and returns are in the 10 percent to 15 percent range for developers/financiers who build out the project and collect PPA revenue.

The biggest argument in favor of pursuing renewable energy development as a business model today is that we are clearly in the elbow of the exponential growth curve, particularly for solar power. Until recently, the future of renewables as a viable business model has always seemed in doubt. There is always some uncertainty in any business venture, but the scale that renewables have reached today, and the growth rates we've seen in recent years, combined with long-term climate change mitigation goals here in the U.S. and globally, weigh heavily in favor of renewables being a highly viable business venture.

Figure 1. *Ernst & Young recently restored the U.S. to its top spot in terms of renewable energy attractiveness*

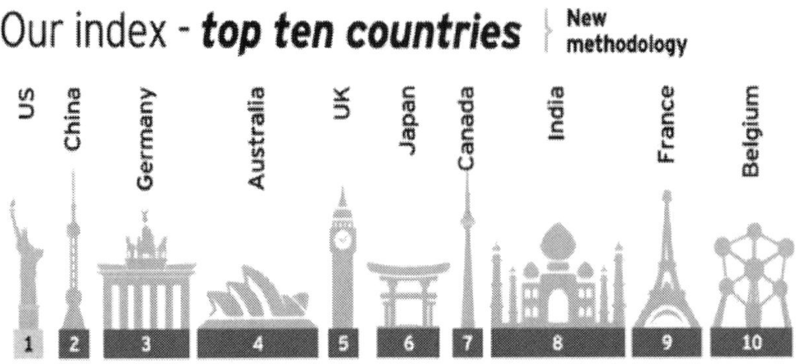

Ernst & Young's annual Renewable Energy Country Attractiveness Index, published annually, restored the U.S. to the top spot in 2013, after crowning China in that spot for the preceding three years. China regained the top spot in 2014, however, so we're seeing an increasing seesawing of these two very big countries in a friendly and beneficial competition for this particular prize.

California is head-and-shoulders above other states in opportunities for renewables, so while it can be a crowded market at times, California is the best state for renewables in one of the two best countries for renewables in the world. Governor Brown instilled some confidence in this market when he called for an increase of the state's renewable electricity standard from one-third renewables by 2020 to one-half by 2030 (and 40% by 2025). These facts should produce some confidence in the business model I've described here for landowners looking to increase profits from their land.

Chapter 23

How to invest in the driverless car future

Driverless cars are the future. This we know. What is not clear, however, is when that future will arrive. The most recent entrant into this exciting field is Cruise Automation, a startup based in San Francisco.

Cruise Automation is testing its aftermarket automated highway driving kit in early 2015. For those who drive an Audi A4 or S4 and want to fork out $10,000, you could be one of the first to be able to read a book or safely check email while driving on the highway without paying any attention to where you're going.

Cruise is a very young company, founded in late 2013, so we should be a bit skeptical of its claims until a bit longer track record has been established.

Other companies are, however, making similarly bold announcements. Tesla's Elon Musk stated in early June: "I am confident that in less than a year you will be able to go from highway on-ramp to highway exits without touching any controls." This is a far cry from fully automated driving but it's a major improvement nonetheless. I look forward to taking long road trips and napping liberally as my car transports my snoring self safely down the road at 75 miles an hour.

Google, another major player in the driverless car arena, surprised a lot of people by shifting its approach to cars with no way for humans to do any driving. Rather than focus on humans as co-pilots in normal cars fitted with sensors, Google's revamped approach to the driverless car future now focuses more modestly on specially designed buggy cars that have very limited human controls. This is because Google has found that human co-pilots test driving its fleet of Lexus SUVs outfitted with sensors and automated controls transformed very quickly, from disbelief and suspicion that computers could drive a car safely, to overconfidence in the computer's ability - and thus a tendency to check out as a co-driver. This, Google believes, isn't a safe option for cars that are intended to have an alert human co-pilot ready to step in if needed.

Google's new specially designed cars are like marshmallows on wheels, complete with a cute face on the front. *Politico*'s Ben White expressed his concern about Google's new approach: "Does no one else get that if you were going to design a soulless killing machine it would look EXACTLY like this?"

Anyway, even if driverless cars are still years away, as may well be the case (particularly for fully automated cars), it is certainly not too early to think about how we can profit from this utopian future.

I do want to first mention the possible dystopian future. IEEE projects that up to 75% of vehicles in the U.S. will be fully automated by 2040, so what I'm suggesting isn't entirely a pipe dream. Some have already expressed concerns about a potential increase in driving from driverless cars making it so easy to take those long road trips, and thus higher greenhouse gas emissions and other pollution. We can't ignore this potential, but my feeling is that any increase in driving by those fortunate enough to have a driverless car will be far outweighed by the increased fuel efficiency that comes with non-human drivers at the helm. And when we consider the very substantial reduction in traffic accidents that will likely occur by getting human drivers out of the driver's seat, it seems pretty clear that the net benefits of driverless cars will be positive.

What follows are my suggestions on how to benefit financially from this major new trend, following up on my recent article looking at how to invest in the renewable energy future. As in my previous piece, my suggestions here are not to be considered any kind of official financial advice. Rather, consider them peer-to-peer tips. And I may be entirely wrong.

Google (GOOG) is perhaps the most obvious company to invest in when it comes to driverless cars. A major benefit of investing in Google is that it comes with a wide host of other technologies and revenue streams, diversifying the risk of a focus solely on driverless cars. Many analysts think Google has plenty of room to grow still. And if the driverless car future takes off in the next few years, it's very likely that Google will be a major player. Google has announced plans to have 100 of its cute little driverless cars on roads in 2015. Is the future here yet? Full disclosure: I just bought some shares of Google in the course of my researching this article.

Tesla (TSLA) is another good - albeit risky - option for investing in the driverless car future simply because Musk and co. have demonstrated their ability to execute and deliver very impressive products. Musk, as mentioned above, has projected automated highway driving in Tesla vehicles by the end of the year and, pending legal issues in various states, it looks promising for Tesla's vehicles to have additional automated driving features steadily added in new models or via software updates. (I haven't focused on legal issues here, but suffice it to say, California and a handful of other states have recently passed laws allowing automated driving to some degree and it seems likely that other states will follow suit before long). Full disclosure: I own some Tesla stock.

Nissan (NSANY) is the most advanced of the majors when it comes to electric vehicles, based on its increasingly popular (and so ugly it's almost cute) electric vehicle, the Leaf. Nissan has announced its intent to offer driverless cars by 2020. This is still a ways out, and will likely be delayed, but, again, I'm talking about *investing* in this article, not trading. And due to its vehicle diversity and size, investing in Nissan is a lot less risky than investing in Tesla.

Investing in lithium stocks is a good way to take advantage of the coming boom in electric vehicles. Driverless cars are related, but of course independent of the propulsion technology used in the vehicle. However, it seems to me that electric vehicles are a natural fit for automated driving features because people who purchase electric vehicles are more drawn to new technology already, and it should be an easier sell to convince people to give the automated driving option a whirl. The lithium ETF LIT is, accordingly, a reasonable bet on the future of driverless cars as well as electric vehicles and energy storage more generally. Full disclosure: I own some LIT shares.

I haven't been able to find any Exchange-Traded Funds (ETFs) for the driverless car future. ETFs are a good way to mitigate risk because they enable investments in a basket of related stocks. It's quite likely that one or more ETFs for the driverless car future will appear in the next year or so and it might behoove farsighted investors to wait until that happens before diving into this space.

Thinking a bit more broadly about how society will be transformed by automated vehicles, it seems that shorting certain companies or sectors may also be a good way to invest in this future. Shorting is inherently risky because there's no guarantee that a given stock will ever go down in price, thus allowing the shorter to recover their money or make some money. However, shorting is a time-honored investment option and it shouldn't be overlooked.

If automated driving does take off, it's quite likely that accident rates will dip sharply because computers are so much better at driving than fallible humans. This gives rise to the possibility of shorting insurance companies in the long-term. Or shorting airlines as people increasingly turn to their automated cars for making longer trips. These last ideas are very speculative so please take a big grain of salt as you consider my suggestions.

In all likelihood, the driverless car future will take many years - at least a couple of decades - to materialize in a way that really transforms our society. But the point of investing is to identify long-term significant trends—big waves that you can ride a long way—and get ahead of the market.

Epilogue

Back from the future: life in 2040

It turns out that some of what I wrote back in 2015 wasn't entirely wrong. Some of it was actually pretty accurate. Not everything of course, but the big picture stuff was surprisingly spot on (pat self on back). I'm no psychohistorian, a la Asimov's *Foundation* series, and I don't think we can mathematize predictions very well at all. But what I did attempt to do in my book was to identify a few robust trends that seemed likely to continue into the future. Specifically, Swanson's Law and Kurzweil's Law of Accelerating Returns provided a scaffolding for seeing the future to some degree.

Anyway, enough self-congratulation. I'm no spring chicken anymore: almost seventy. But I feel like I'm 40 and I'm told I don't look a day over 68. Technology can do some things right and skin enhancement techniques, along with other rejuvenation technologies, are getting pretty dang good.

I'm actually writing this while I drive to Banff National Park for the annual Mountain Film Festival there. Well, I'm not really driving. My car is driving itself, of course, and I'm really enjoying the views of the Canadian Rockies as I cruise through southern British Columbia on my way to Alberta's beautiful national park.

I've turned off all non-emergency outside communications so it's just me, my laptop and my electric car traveling through nature at this point. My wife and kids know how to reach me if they really need to but for now non-distraction is the name of the game. We have to battle in today's tech ping buzz whiz world for some privacy with our own thoughts.

There are no wind turbines up here in the Rockies but I passed a ton of them on my drive through California, Oregon and Idaho on my way from Santa Barbara. They're beautiful things, these modern turbines. Even the older ones have a gorgeous swept back curved-blade design. Most newer ones include tubercles, like we see on humpback whale fins, as an additional design feature to reduce turbulence and thus increase efficiency.

I saw a lot of solar panels too, of course. They're pretty ubiquitous nowadays. I used to be able to drive from California to Washington State up the whole West Coast, in the twenty-teens, and see almost no solar panels or turbines. That's changed. Big time. There are a ton of medium-scale solar installations visible, particularly along highway medians and in fields not far from the highways. And where there are rail lines there are usually solar canopies over the rail lines taking advantage of all that open space.

My car's solar panels are a gorgeous dark blue and you wouldn't even know they were part of the car unless someone told you. They don't get me too many miles of driving, even with today's radical efficiency of 70 percent. But the on-car panels help run my electronics and air conditioning.

Most highways now include charging stations every few miles so that cars can automatically pull in and charge for the fifteen minutes or so it takes for a 400-mile charge. Since the whole process is automatic anyway it's nice to take a little bathroom break and stretch my legs. And because most of these stations are powered by solar and wind power power I don't even feel guilty that I drive so much nowadays. For most trips around North America I just tell my car to take me there, pack my overnight bag, nap, work, talk on the phone, enjoy the view, and lo and behold in less than a day I can get to just about any place I'd want to go on this big old continent.

High-speed trains and hyperloops are increasingly common too, and most are solar-powered. Given these new mass transit options, I feel a little guilty taking my car on these long trips sometimes—but not that guilty because my car is almost entirely powered by renewables so the impacts are really pretty minimal.

When the large majority of cars on highways are automated we go a lot faster than we used to. My average speed on this trip to Banff has been over one hundred miles per hour. Ironically, accidents are far less frequent than they used to be. Even now, however, laws forbid drinking while driving and that's probably pretty smart. There are occasions when the car has to ask for human intervention, but it's been years, now that I'm thinking about it, that my car has asked me to intervene on anything. So maybe they'll change that no drinking policy before too long. From now on I'll keep some champagne in my car in anticipation.

It's a shame that most of the glaciers in this part of the world have melted. But at least we were able to turn the trend on greenhouse gases in the late twenty-teens. 2019 was in fact the peak year for global emissions and even though China and India's standard of living is getting closer to ours here in North America, we were able to keep emissions low because those countries, and other major developing countries like Indonesia, Brazil and Nigeria, took advantage of the solar, wind and EV revolutions. Most formerly developing countries were able to build out largely distributed energy grids that rely now mostly on solar, wind, biomass, geothermal and energy storage. Natural gas is still used in many parts of the world but even that is in heavy decline as the solar revolution continues to work its magic.

The U.S. grid is now about 80% renewable energy and petroleum use is also down dramatically at 60% below the peak year of 2017. EVs caught on big-time after the historic Tesla Model 3 and Chevy Bolt came out in 2017. Very few gave those companies much credence at the time in terms of their cars' potential to be truly transformative. But we now have the proof in the pudding: those two cars became the trendsetters for every major car company and every manufacturer stepped up by 2020 with their own affordable EV with over 200 miles of range. From then on, we quickly saw EVs become the majority of new cars produced and sold around the world.

EV range doubled to 400 miles by 2025 and it's stayed at that level since then for most cars because further improvements have focused on battery cost reductions and weight reductions. Cars are now half as heavy as they used to be, even with the enhanced range, and batteries are far more energy dense than they were in the late twenty-teens.

Unfortunately, practically free energy hasn't solved all of the world's problems yet. We are well on our way to bringing atmospheric carbon emissions down to 350 parts per million and I'm optimistic that the climate will stabilize before too long. Storms are increasingly crazy and the damage wreaked by hurricanes, tornadoes and major storms has skyrocketed. That's a topic for a different day, but at least we know we're on our way to mitigating these issues with our dramatic reduction in global emissions. Now all the talk is about major geo-engineering projects to scrub the carbon dioxide from the atmosphere.

Almost free energy hasn't done much yet to solve the world's political problems. The Middle East is even more dangerous than it was in the twenty-teens because much of the region's revenue dried up in the 2020s with big reductions in oil demand. The price of oil plummeted again in 2021, a few years after the plummets of

2014 and 2008. The 2008 plummet was due to a global economic crisis and prices quickly rebounded after the financial crisis was averted. Prices dropped so fast in 2014 because of the fracking revolution and declining demand in big economies like China, but by 2016 they'd gone back to over $100 a barrel because the new oil being produced couldn't make up for the declines in conventional fields. The 2021 drop, however, was lasting, and it was induced mainly by the phenomenal success of the new crop of EVs with 200+ miles of range.

It took a few years for this trend to become apparent but by 2020 most observers realized the revolution was here to stay and that oil demand was going to decline and keep on declining as EV sales skyrocketed. It's been a fun ride for most of us, but the ongoing conflicts in the Middle East and other oil-producing nations do worry me. Russia has finally accepted its former great power status and Russia, China and India use their permanent veto power at the United Nations sparingly.

China, India, Brazil, Japan, the EU and the U.S. are now generally viewed as co-equals in a truly multipolar world. Even the African Union is gaining significant respect and power in today's international system. Who would have thought that Africa, for so long a political and economic basket case, would eventually get its act together?

Thankfully, the world's powers saw the multi-polar world clearly enough in the twenty-twenties to make a big shift toward a more empowered international system. This has helped to reduce conflict between the major powers for the most part, but there is still a bit of nostalgia for the "good old days" when the U.S. was relied upon as an (erratic) arbiter of international order.

I think that's it for now. I could ramble on further (and I'm told that I do tend to ramble sometimes) but let's end it there. I'm going to kick back and enjoy the view. I love mountains.

About the author

Tam Hunt is a lawyer and energy policy expert based in Santa Barbara, California, and Hilo, Hawaii. He is the founder and owner of Community Renewable Solutions LLC, a consulting and law firm specializing in community-scale renewable energy. He was a visiting lecturer at UC Santa Barbara's Bren School of Environmental Science & Management from 2007-2014. He is also a contributor to a number of online energy publications, including GreenTechMedia.com and RenewableEnergyWorld.com, where a number of the chapters in this book first appeared in different forms.

Made in the USA
Columbia, SC
03 December 2017